HUMAN AND MACHINE CONSCIOUSNESS

Human and Machine Consciousness

David Gamez

https://www.openbookpublishers.com

© 2018 David Gamez

This work is licensed under a Creative Commons Attribution 4.0 International license (CC BY 4.0). This license allows you to share, copy, distribute and transmit the work; to adapt the work and to make commercial use of the work providing attribution is made to the authors (but not in any way that suggests that they endorse you or your use of the work). Attribution should include the following information:

David Gamez. *Human and Machine Consciousness*. Cambridge, UK: Open Book Publishers, 2018. https://doi.org/10.11647/OBP.0107

In order to access detailed and updated information on the license, please visit https://www.openbookpublishers.com/product/545#copyright

Further details about CC BY licenses are available at http://creativecommons.org/licenses/by/4.0/

All external links were active at the time of publication unless otherwise stated and have been archived via the Internet Archive Wayback Machine at https://archive.org/web

Digital material and resources associated with this volume are available at https://www.openbookpublishers.com/product/545#resources

Every effort has been made to identify and contact copyright holders and any omission or error will be corrected if notification is made to the publisher.

ISBN Paperback: 978-1-78374-298-1
ISBN Hardback: 978-1-78374-299-8
ISBN Digital (PDF): 978-1-78374-300-1
ISBN Digital ebook (epub): 978-1-78374-301-8
ISBN Digital ebook (mobi): 978-1-78374-302-5
DOI: 10.11647/OBP.0107

Cover image: Stereogram created by David Gamez with data from Anderson Winkler (https://brainder.org/research/brain-for-blender/) licensed under CC BY-SA 3.0.

All paper used by Open Book Publishers is SFI (Sustainable Forestry Initiative), PEFC (Programme for the Endorsement of Forest Certification Schemes) and Forest Stewardship Council(r)(FSC(r) certified.

Printed in the United Kingdom, United States, and Australia by Lightning Source for Open Book Publishers (Cambridge, UK)

This book is dedicated to the first artificial system that understands it.

A flash, a mantling, and the ferment rises,
Thus, in this moment, hope materializes,
A mighty project may at first seem mad,
But now we laugh, the ways of chance forseeing:
A thinker then, in mind's deep wonder clad,
May give at last a thinking brain its being.
[...]
Now chimes the glass, a note of sweetest strength,
It clouds, it clears, my utmost hope it proves,
For there my longing eyes behold at length
A dapper form, that lives and breathes and moves.
My mannikin! What can the world ask more?
The mystery is brought to light of day.
Now comes the whisper we are waiting for:
He forms his speech, has clear-cut words to say.

Goethe, *Faust*

Acknowledgements

I am extremely grateful to Barry Cooper and the John Templeton Foundation for supporting this work (Project ID 15619: 'Mind, Mechanism and Mathematics: Turing Centenary Research Project'). This grant gave me the time that I needed to sit down and write this book.

I have really appreciated the help of Anil Seth, who supported my application for a Turing Fellowship and was very welcoming during my time at the University of Sussex. I am also grateful to the Sackler Centre for Consciousness Science and the Department of Informatics at the University of Sussex for giving me a place to work. I greatly enjoyed conversations about consciousness with my colleagues at Sussex.

I would also like to thank Owen Holland, whose CRONOS project started my work on human and machine consciousness, and the reviewers of this book, who had many helpful suggestions. I owe a warm debt of gratitude to my parents, Alejandro and Penny Gamez, who have always given me a great deal of support and encouragement.

Contents

List of Illustrations 1

1. Introduction 3
2. The Emergence of the Concept of Consciousness 9
3. The Philosophy and Science of Consciousness 33
4. The Measurement of Consciousness 43
5. From Correlates to Theories of Consciousness 69
6. Physical Theories of Consciousness 85
7. Information Theories of Consciousness 93
8. Computation Theories of Consciousness 103
9. Predictions and Deductions about Consciousness 113
10. Modification and Enhancement of Consciousness 125
11. Machine Consciousness 135
12. Conclusion 149

Appendix: Definitions, Assumptions, Lemmas and Constraints 159
Endnotes 165
Bibliography 201
Index 219

List of Illustrations

All images are © David Gamez, CC BY 4.0.

2.1.	Visual representation of a bubble of perception.	12
2.2.	The presence of an invisible god explains regularities in the visible world.	14
2.3.	Colour illusion.	17
2.4.	Primary and secondary qualities.	19
2.5.	The relationship between a bubble of experience and a brain.	21
2.6.	Interpretation of physical objects as black boxes.	23
2.7.	The relationship between a bubble of experience and an invisible physical brain.	25
2.8.	The emergence of the concept of consciousness.	28
3.1.	The use of imagination to solve a scientific problem.	35
3.2.	Imagination cannot be used to understand the relationship between consciousness and the invisible physical world.	38
3.3.	Learnt association between consciously experienced brain activity and the sensation of an ice cube.	39
4.1.	Problem of colour inversion.	51
4.2.	Some of the definitions and assumptions that are required for scientific experiments on consciousness.	53
4.3.	The relationship between macro- and micro-scale e-causal events.	58
4.4.	Assumptions about the relationship between CC sets, consciousness and first-person reports.	60
5.1.	The measurement of an elephant's height in a scientist's bubble of experience.	70
5.2.	Theory of consciousness (c-theory).	79

7.1. Information c-theory.	97
8.1. Soap bubble computer.	104
9.1. Testing a c-theory's prediction about a conscious state.	114
9.2. Testing a c-theory's prediction about a physical state.	115
9.3. Deduction of the conscious state of a bat.	119
10.1. Modifications of a bubble of experience.	128
10.2. A reliable c-theory is used to realize a desired state of consciousness.	129
11.1. A reliable c-theory is used to build a MC4 machine.	138
11.2. A reliable c-theory is used to deduce the consciousness of an artificial system.	139

1. Introduction

Consciousness is extremely important to us. Without consciousness, there is just nothingness, death, night. It is a crime to kill a person who is potentially conscious. Permanently unconscious people are left to die. Religious people face death with hope because they believe that their conscious souls will break free from their physical bodies.

We know next to nothing about consciousness and its relationship to the physical world. The science of consciousness is mired in philosophical problems. We can only guess about the consciousness of coma patients, infants and animals. We have no idea about the consciousness of artificial systems.

This book neutralizes the philosophical problems with consciousness and clears the way for scientific research. It explains how we can develop mathematical theories that can make believable predictions about consciousness.

The first obstacles that need to be overcome are the metaphysical theories of consciousness. Some people claim that consciousness is a separate substance; other people believe that it is identical to the physical world. These theories generate endless debates and it is very difficult to prove or refute them. This book eliminates some of these theories and suspends judgement about the rest.

The next obstacle is the hard problem of consciousness. This typically appears when people try and fail to imagine how colourful conscious sensations are related to the colourless world of modern physics. This book breaks the hard problem of consciousness down into a pseudo problem, a difficult problem and a set of brute regularities.

Some problems with consciousness cannot be solved. For example, we cannot prove that a person is conscious. These problems affect our ability to measure consciousness through first-person reports. This book neutralizes these problems by making *assumptions*. The results from the science of consciousness can then be considered to be true *given these assumptions*.

When these obstacles have been overcome the scientific study of consciousness becomes straightforward. We can measure consciousness, measure the physical world and look for mathematical relationships between these measurements. We can use artificial intelligence to discover mathematical theories of consciousness.

Eventually we will discover mathematical theories that map between states of consciousness and states of the physical world. We will use these theories to make believable predictions about the consciousness of infants, animals and robots. We will measure the consciousness of brain-damaged patients. We will build conscious machines, repair damaged consciousnesses and create designer states of consciousness.

The scientific study of consciousness is clarified by this book. As you read it the philosophical problems will dissolve and you will gain a clear vision of consciousness research. You will no longer worry about whether consciousness is a separate substance. You will not be troubled by a desire to reduce consciousness to particles or forces. You will understand that a scientific theory of consciousness is a mathematical relationship between a formal description of consciousness and a formal description of the physical world.

This book starts with a definition of consciousness. In daily life we treat colour, sound and smell as objective properties of the world. Over the last three hundred years science has developed a series of interpretations of the world that have stripped objects of their sensory properties. Apples used to be red and tasty; now physical apples are colourless collections of jigging atoms, probability distributions of wave-particles. The physical world has become invisible. When science eliminated sensory properties from the physical world it was necessary to find a way of grouping, describing and explaining the colours, sounds and smells that we continued to encounter in daily life. We solved this problem by inventing the modern concept of consciousness. 'Consciousness' is a name for the sensory properties that were removed from the physical world by modern science.

The next chapter examines some 'hard' problems with consciousness. First, it is impossible to imagine the relationship between consciousness and the invisible physical world. Second, we find it difficult to imagine the connection between conscious experiences of brain activity and other conscious experiences. Third, there are brute regularities between consciousness and the physical world that cannot be broken down or further explained. None of these problems are unique to consciousness research. They can also be found in physics and they do not affect our ability to study consciousness scientifically. We can measure consciousness, measure the physical world and look for mathematical relationships between these measurements.

Scientists measure consciousness through first-person reports, which raises problems about the reliability of these reports, the possibility of non-reportable consciousness and the causal closure of the physical world. The fourth chapter addresses these issues by making assumptions that explain how consciousness can be measured. First, we need to identify the systems that we believe are conscious. Then we need to make other assumptions to ensure that consciousness can be accurately measured in these systems.

The fifth chapter explains how we can develop mathematical theories of the relationship between consciousness and the physical world. Scientists have carried out pilot studies that have looked for correlations between consciousness and brain activity. We are now starting to create compact mathematical theories that can map between physical and conscious states. Computers could be used to discover these theories automatically.

Chapter 6 discusses theories that link consciousness to patterns in physical materials—for example, electromagnetic waves or neuron firing patterns. With physical theories the materials in which the patterns occur are critical—if the same patterns occur in different materials, they are not claimed to be linked to consciousness. Physical theories of consciousness are similar to scientific theories in physics, chemistry and biology.

Some people have claimed that information patterns are linked to consciousness, regardless of whether they occur in a brain, a computer or a pile of sand. The seventh chapter shows that this approach fails because information is not a property of the physical world and any given information pattern can be extracted from both the conscious and unconscious brain. Information theories of consciousness should be reinterpreted as physical theories of consciousness.

Other people believe that consciousness is linked to the execution of computations. They claim that some computations are linked to consciousness regardless of whether they are executing in a brain or a digital computer. Chapter 8 argues that computations cannot be linked to consciousness because computing is a subjective use that we make of the world. Computation theories of consciousness should be reinterpreted as physical theories of consciousness.

Chapter 9 explains how theories of consciousness can be experimentally tested. This can only be done on systems that we assume are conscious, such as normally functioning adult human brains. We can also use our theories of consciousness to make deductions about the consciousness of brain-damaged people, animals and robots. These deductions cannot be verified because we cannot measure the consciousness of these systems.

When we have discovered a reliable theory of consciousness we will be able to use it to modify and enhance our consciousness. For example, we could change the shape of our conscious body or increase our level of consciousness. Chapter 10 explains how we can use a theory of consciousness to identify the physical state that is linked to a desired conscious state. If we could realize this physical state in our brains, we would experience the desired conscious state. It will be many years before this will become technologically possible.

The eleventh chapter suggests how a reliable theory of consciousness could be used to create conscious machines and make believable deductions about the consciousness of artificial systems. Silicon brain implants and consciousness uploading are interpreted as forms of machine consciousness, and the chapter discusses whether conscious machines could threaten human existence and how they should be ethically treated.

The conclusion summarises the book, highlights its limitations and suggests future directions of research. The appendix lists the definitions, assumptions, lemmas and constraints.

The main text of this book is short and self-contained and can be read through without referring to the endnotes or bibliography. The endnotes contain more detailed discussions of individual points and full references to the scientific and philosophical literature.

2. The Emergence of the Concept of Consciousness

2.1 Naive Realism

I am immersed in a colourful moving noisy tasty smelly painful spatially and temporally extended stream of things. During a nuclear explosion I see a grey mushroom cloud, hear a detonation, feel heat, touch wind and taste synthetic strawberry bubblegum in my mouth. I do not *infer* the presence of these things—they are just there before me as the world at this place and time seen from my perspective.

When Cro-Magnon man peered out of his cave he saw a bright pattern of green leaves, heard a river and tasted sweet-tart berries in his mouth. The green of the leaves was present to him, framing the entrance to his cave, just as the river was crashing and roaring to his left. No complicated theories about consciousness troubled Cro-Magnon man: the world was simply present to him. In this idealised naive and simple time people simply saw the world, unclouded by theories of perception.

When a child opens its eyes it does not see a collection of qualia[1] or conscious representations: just a red balloon ascending into the warm summer sky.

Most modern adults most of the time have a direct relationship with the world around them. We are immersed in a world of colourful moving noisy tasty smelly things. As we slog through our workaday lives we are not *philosophizing*—the blue of my computer screen is the colour of an object in the world; the tinny speaker sound is part of the world. We go outside and see cold grey skies and are lashed by cold lashing rain.

For me at least, the colourful cheerful world is the most important thing there is. I long to drink in more of the visible audible tasty moving world. What I hope for in any afterlife is that some kind of a world will continue, ideally in a reasonably pleasant way. While one can make abstract ethical points about the value of life, its real value for me is this immersion in a sensuous world.

This relationship with the world is often called *naive realism*: an interpretation of perception in which we directly see the world and the world is as we see it. However, there is nothing *naive* or *realistic* in our everyday encounters with the world—'naive realism' is a convenient label that we use to contrast our everyday immersion in the world with other theories of perception.

I am standing in my sitting room staring dully through dirty net curtains at nothing in the street outside. I cannot see the body of my aunt. It is out there in the garage. I walk into the garage and open the blue plastic sack. Now I can see the body of my aunt.

When I look at my aunt's body it appears *as* three-dimensional, although I can only see part of it at one time. From one perspective I can see my aunt's grey lips and clouded eyes, but I cannot see her whole head or body. I have to move relative to her body to see her thin grey hair and the matted dried blood on the back of her head.

My aunt's body changes independently of my interactions with it. Each time I return to the garage I observe subtle changes in colour as her body decays. Her body has an *objective* existence that can be systematically probed in different ways. I can perform chemical tests; I can measure its hardness and weight.

Other people cannot see the body of my aunt. The police cannot see it. Uncle Henry, on holiday in Tahiti, is staring at the gyrating buttocks of a young woman in a grass skirt. He is not looking at the body of my aunt.

Naive realism is not simultaneous and all-embracing access to every object in existence. We see a small number of the world's objects from *one perspective*. Objects have an *independent existence* that enables them to be perceived by other people. Different people *see different things*. We can perceive the *same object* on *multiple occasions*. Objects can be in *different states at different times*.

2. The Emergence of the Concept of Consciousness

In our naively realistic encounters with the world we use the language of perception to indicate those things and those aspects of things that are present to us and to acknowledge that objects continue to exist when they are not being perceived. Instead of saying that my aunt's body *is there*, I talk about *perceiving* my aunt's body to indicate that it is currently present to me. Uncle Henry is not perceiving her body: it is not present to him in Tahiti.

Perception is similar to a bubble that we 'carry around' with us that contains the objects that are currently present to us. I will call this a *bubble of perception*. We are immersed in our bubbles of perception. When an object appears in my bubble of perception I see it from a perspective that is centred on my body.[2]

A visual representation of a bubble of perception is shown in Figure 2.1b. This is inaccurate because it shows the person's body from a third-person perspective, whereas we experience our bubbles of perception from the inside—we look out from our bodies onto the world. This illustration has the further limitation that it only shows the visual aspect of a bubble of perception. Bubbles of perception also include tastes, sounds, smells, body sensations and emotional states.

In naive realism objects have the properties that we perceive them to have. The plastic sack *is blue*; my aunt's body *is cold*; her clothes *have a mothball and urine odour*. Objects have these properties independently of whether they are inside or outside a bubble of perception. The plastic sack continues to be blue when it is in the garage and not being perceived by anyone (Figure 2.1a).

I sit in the kitchen and imagine my aunt's body in the garage. Now the contents of the sack are fleeting and unstable, colours are washed out and the smell of moth balls and urine is not present. I dream of my aunt's body. This is more vivid than imagination, but my aunt's face changes from moment to moment, and it is difficult to inspect details and maintain consistency over time. I go for a walk in the forest and eat a mushroom. One hour later my aunt rises from the ground before me: her eyes are dark geometric spirals; her hair is a writhing mass of white maggots.

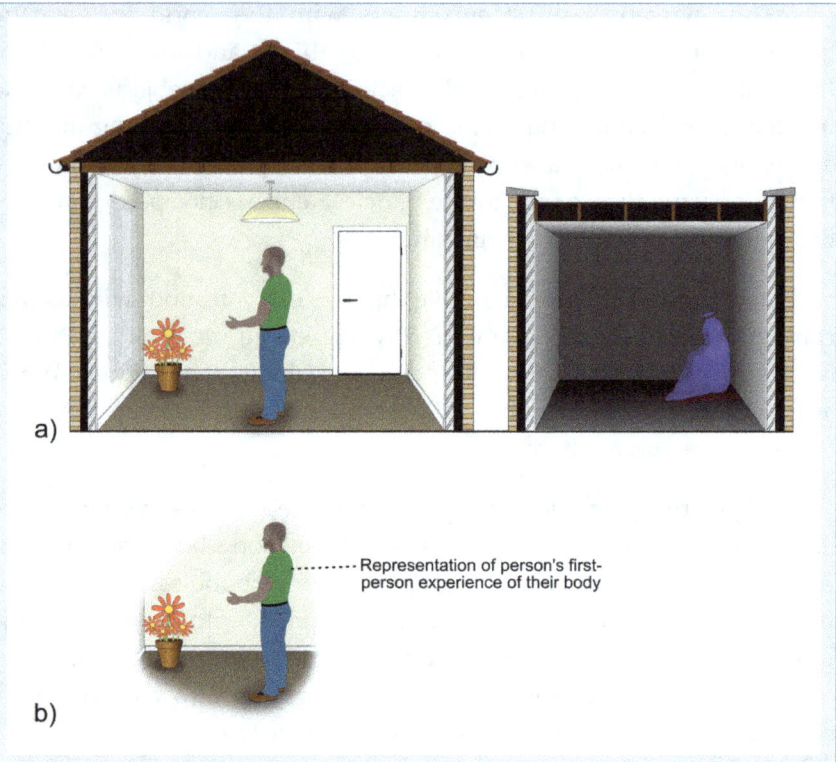

Figure 2.1. Visual representation of a bubble of perception. a) Domestic scene. In naive realism the sack in the garage continues to be blue when no-one is looking at it. b) A visual representation of a bubble of perception. This uses a third-person perspective to represent our sense of inhabiting a body and looking out at a world. Although this is substantially different from an actual bubble of perception, which we experience from *inside* our bodies, it is the best way that I have found of depicting a bubble of perception. Image © David Gamez, CC BY 4.0.

We no longer believe that imagined, dreamt or hallucinated objects are objectively present in a second spiritual world. It no longer makes sense to say that we *perceive* imagined, dreamt or hallucinated objects. This is particularly true now that perception is associated with theories about electromagnetic waves, sound vibrations, and so on. To address this issue I will replace 'bubble of perception' with the more inclusive term 'bubble of experience', and distinguish between two types of bubble of experience:

- *Online* bubbles of experience are connected to the world: their states change in response to changes in the world and detailed information about the world can be accessed on demand.

They typically have vivid colours, clear sounds, strong odours and intense body sensations. In an online bubble of experience objects are stable, we can view the same object on multiple occasions and people generally agree on an object's properties. We are immersed in online bubbles of experience when we perceive and interact with the world.

- *Offline* bubbles of experience are not connected to the current environment, although they might correspond to past or future states. They are often unstable, low resolution and low intensity. Colours are washed out; smells, tastes and body sensations are rarely present. Offline bubbles of experience are typically weakly perceptual—we cannot interact with objects in a systematic way, and it can be difficult to repeatedly view the same object from multiple perspectives or to examine small details. People typically do not agree about the objects that they encounter in offline bubbles of experience. We are immersed in offline bubbles of experience when we dream, remember, hallucinate and imagine.

A bubble of experience can have a mixture of online and offline contents. When I hallucinated my aunt the forest was an online component of my bubble of experience; the aunt and maggots were offline.[3]

2.2 Invisible Explanations

The flowers in my living room appear in my online bubble of experience on multiple occasions. I can see them from multiple perspectives and uncover more of their properties. They appear in other people's bubbles of experience. The flowers are part of an independent world, which is often called the *physical world*.

The physical world has regularities. If I throw a pig out of a window, its pink colour and screams move together and its rate of acceleration can be calculated using a simple equation. If I mix one part glycerine with three parts nitric acid, I obtain an explosive mixture that can alleviate angina.

We explain these regularities by postulating the existence of invisible objects and properties in the physical world. These do not appear in our bubbles of experience—we believe in their existence because they improve our ability to make predictions about objects in our bubbles of experience.

X-rays are invisible waves that were posited to explain the appearance of patterns on photographic plates. These patterns can easily be explained if there is a form of radiation that cannot be perceived with the human eye. Our belief in X-rays was strengthened by the development of other methods for detecting them. Only the *effects* of X-rays appear in our bubbles of experience—the rays themselves are invisible.

Visible and invisible gods are often used to explain regularities in our bubbles of experience. A statue of Tlaloc might be considered to be Tlaloc himself, something that Tlaloc inhabits to some extent or just a representation of Tlaloc. Sometimes the Judeo-Christian god is depicted as a beardy bloke floating in the clouds; more often he is assumed to be invisible.

Prayers, sacrifices and moral rectitude encourage the gods to bestow rain, fertility and a good harvest on their virtuous subjects (see Figure 2.2). Murder, incest and eating prawns anger the gods, who inflict earthquakes, floods and infertility on people who stray from the path of righteousness.

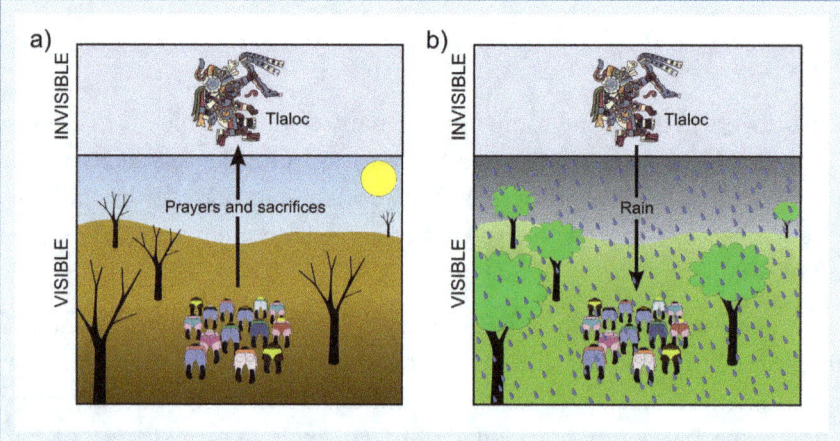

Figure 2.2. The presence of an invisible god explains regularities in the visible world. a) Worshippers of Tlaloc offer up sacrifices and prayers for rain. b) The psychology and actions of the invisible god explain the appearance of the rain. Image © David Gamez, CC BY.

Early astronomers explained the regular movements of the heavenly bodies by claiming that they are embedded in concentric crystalline spheres. These spheres were invisible to human observers on Earth, but they probably believed that they could have touched them if they could have reached them.

Newton explained the movements of the heavenly bodies by claiming that they exert an invisible gravitational force on each other, whose strength is given by a simple equation. Newton could not explain *how* masses attract each other at a distance—at best he could point to magnetism as an example of a similar force. However, the invisible gravitational force, along with the equations describing it, made good predictions about the movements of the heavenly bodies, and so it became an accepted part of the physical world. While we can observe the *effects* of gravity in our bubbles of experience—a feeling of heaviness, movement of objects towards the Earth—gravity itself is invisible.

The ancient atomists hypothesized that the world is composed of invisible entities called atoms. They used the movements, swerves and interactions of the atoms to explain the visible properties of the world.[4] This view was revived in the seventeenth century and later used to explain phenomena, such as the pressure and temperature of a gas. Although our theories about elementary particles have been substantially revised, atomism continues to play an important role in our understanding of the physical world.

Atoms and their constituent particles are invisible explanations because they never directly appear in our bubbles of experience. An atom might emit an electromagnetic wave that leads to an experience of red, but we experience the red, not the atom itself. We can generate pictures of atoms using a scanning tunnelling microscope, but these are the result of a complex technological process—not a direct view of the atoms themselves.

Our modern invisible explanations have become increasingly abstract. We now use complex mathematical equations to describe the behaviour of wave-particles and highly folded fields. These invisible explanations can be used to make accurate predictions about the behaviour of objects in our bubbles of experience.

Invisible physical explanations are extremely important to us. For non-religious people the physical world is all there is: a complete understanding of it would be a complete understanding of everything.

Whichever invisible explanations you accept, their common factor is that they are, by definition, *invisible*. They are hypotheses that go beyond our experiences in order to explain and make sense of our experiences. The *effects* of invisible entities appear in our bubbles of experience, never the invisible entities themselves.

2.3 Primary and Secondary Qualities

The particular *bulk, number, figure, and motion of the parts of fire, or snow, are really in them*, whether anyone's senses perceive them or no: and therefore they may be called *real qualities*, because they really exist in those bodies. But *light, heat, whiteness*, or *coldness, are no more really in them, than sickness or pain is in* manna. Take away the sensation of them; let not the eyes see light, or colours, nor the ears hear sounds; let the palate not taste, nor the nose smell, and all colours, tastes, odours, and sounds as they are such particular ideas, vanish and cease, and are reduced to their causes, *i.e.* bulk, figure, and motion of parts.

John Locke, *An Essay Concerning Human Understanding*[5]

Some honey is in my bubble of experience. It feels warm and sticky. It is a dark semi-translucent brown colour. I taste the honey—it is sweet. I put the honey in a box and close the lid. Although the honey is no longer in my bubble of experience, it is natural to assume that it continues to be sweet, warm, sticky, and semi-translucent dark brown in colour.

I pass the honey to Zampano. He tastes the honey. 'Cor blimey stab me vitals,' he says, 'that's some bitter honey.' In *his* bubble of experience the honey is bitter. So is the honey sweet, bitter, both sweet and bitter, or neither when it is outside our bubbles of experience?[6]

The honey changes colour when I put it in different contexts and expose it to light of different colours (see Figure 2.3). What is the colour of the honey in and of itself? What is the colour of the honey when it is outside my bubble of experience? When it is in the dark? When it is viewed by a snake?

These contradictions in an object's properties can be resolved by an account of perception that attributes some properties to the physical world and other properties to the interaction between the physical world and the senses. A good example of this approach is Galileo's and Locke's distinction between primary and secondary qualities.[7] Primary qualities, such as size, shape and movement, are properties of the objects themselves. Secondary qualities, such as colour, smell and sound, arise when the physical world interacts with the senses—they are not properties of physical objects.

The size, shape and movement of honey are primary qualities: properties that honey has regardless of whether it is perceived or not. These properties are intrinsic to all physical objects. The colour and sweetness of honey are secondary qualities that arise when honey interacts with a person's senses. When honey is outside all bubbles of experience it is not sweet, bitter or coloured in any way.

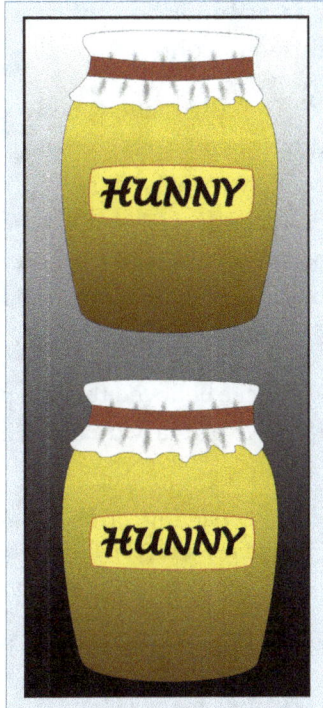

Figure 2.3. Colour illusion. The jars of honey are identical; the shaded background makes the top jar appear to be darker in colour. Image © David Gamez, CC BY 4.0.

Different bodies have different sense organs and interact in different ways with their environment. This explains how the same physical object can produce different secondary qualities in different people. When honey interacts with my senses it produces sensations of warmth, sweetness and a dark semi-translucent brown colour. When it interacts with Zampano's senses it produces coldness, bitterness and a dark semi-translucent orange colour. This account of perception avoids the attribution of contradictory properties to the same physical object. It explains how honey can be perceived as sweet by some people and as bitter by others; why the same patch has different colours in different contexts.[8]

The distinction between primary and secondary qualities was developed in response to the revival of atomism in the seventeenth century. Atoms were hypothesized to be the fundamental constituents of the physical world and primary qualities were properties of the atoms. Interactions between atoms in the environment and atoms in our bodies led to the appearance of secondary qualities, such as redness and sweetness (see Figure 2.4).

Locke believed that the primary qualities in our bubbles of experience *resemble* primary qualities in the physical world:

> [...] the *ideas of primary qualities* of bodies, *are resemblances* of them, and their patterns do really exist in the bodies themselves; but the ideas, *produced* in us *by* these *secondary qualities, have no resemblance* of them at all. There is nothing like our ideas, existing in the bodies themselves. They are in the bodies, we denominate from them, only a power to produce those sensations in us: and what is sweet, blue or warm in idea, is but the certain bulk, figure and motion of the insensible parts in the bodies themselves [...][9]

When I am hugging a moving medium-sized bear in my online bubble of experience there is a moving medium-sized bear in the physical world. According to Locke the size, shape and motion of the bear in my bubble of experience match the size, shape and motion of the bear in the physical world. However, there is no growling sound, brown colour or pungent bear-smell in the physical world. Air vibrations, electromagnetic waves and molecules in the physical world interact with my sense organs to produce the growling sound, brown colour, and pungent bear-smell in my bubble of experience.

The primary qualities of physical objects are perceived through their secondary qualities. We cannot discover the size, shape or motion of an object without perceiving its colour, hearing its sound or touching it. Objects might possess their primary qualities independently of our perception of them, but these primary qualities can only appear in our bubbles of experience when they are clothed in secondary qualities. Physical objects are completely invisible without secondary qualities.

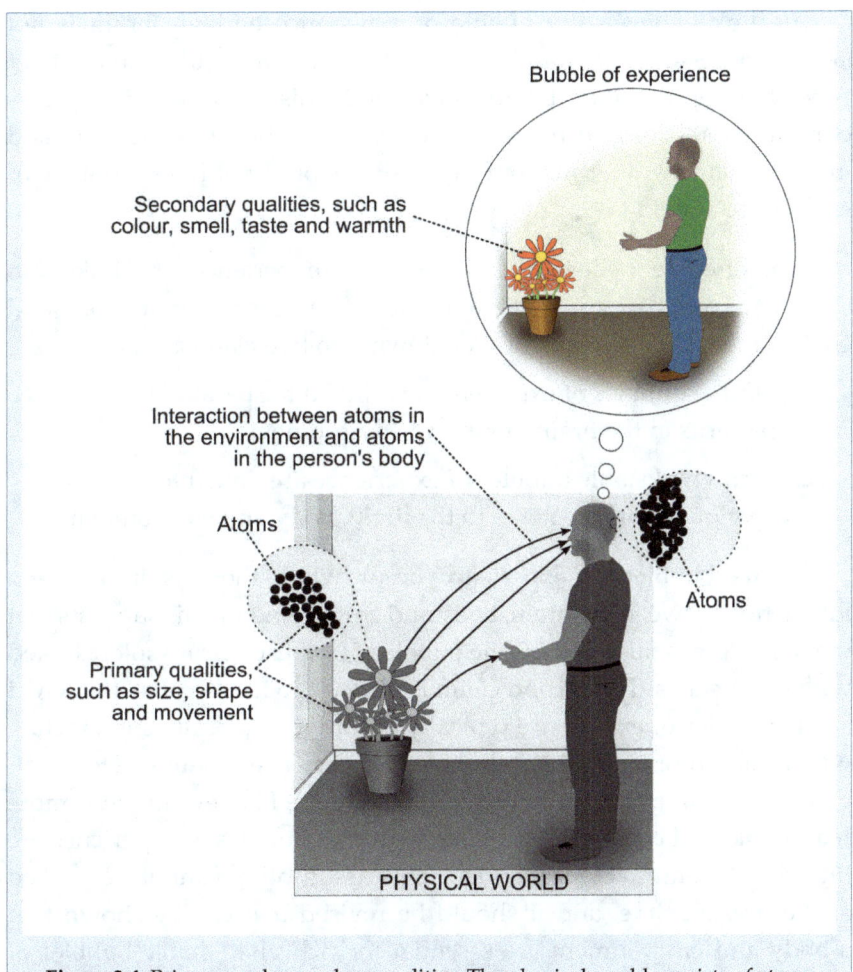

Figure 2.4. Primary and secondary qualities. The physical world consists of atoms with primary qualities, such as size, movement and shape. When atoms interact with a person's sense organs they give rise to secondary qualities, such as colour, smell, taste and warmth, that appear in their bubble of experience. Image © David Gamez, CC BY 4.0.

2.4 Bubbles of Experience and the Brain

When I hit my head my bubble of experience is filled with bright points of light. Stimulation of my brain with electrodes evokes visual, auditory and somatic sensations. Brain damage damages my bubble of experience. My experiences can be altered by changing my brain's chemical state.

The link between my bubble of experience and my brain is not *logically* necessary—it would not be a *contradiction* if a blow to my liver produced bright points of light. However, in this world, with these laws of nature, the strong correlations between my bubble of experience and my brain suggest that without my brain I would not have a bubble of experience at all.

Some people believe that bubbles of experience are linked to spatiotemporal patterns that are distributed across the brain, body and environment.[10] This can be broken down into two claims:

1. Offline bubbles of experience are linked to spatiotemporal patterns in the brain, body and environment.
2. Rich vivid stable bubbles of experience are linked to spatiotemporal patterns in the brain, body and environment.

Offline bubbles of experience occur when there is little or no interaction between the brain, body and environment. This suggests that the first claim is false and offline bubbles of experience are solely linked with brain states. The second claim is difficult to test because rich vivid stable bubbles of experience typically occur when a brain is interacting with its environment (when a bubble of experience is online). However, given everything that we know about the brain, I believe that it is more reasonable and economical to assume that *all* bubbles of experience are linked with brain activity alone.[11] This assumption cannot be proved at the present time, and it should be revised if it can be shown that a body and environment are essential for rich vivid stable bubbles of experience.[12]

When we are immersed in an online bubble of experience our bodies are interacting with our environment and our sense organs are passing streams of spikes[13] down our nerves and changing the states of our brains (see Figure 2.5). When we are immersed in an offline bubble of experience the states of our brains are changing independently of our body and environment. In both cases I will assume that our bubbles of experience are only linked with states of our brains.

In the previous section it was suggested that primary qualities are perceived through secondary qualities and that the primary qualities in

2. The Emergence of the Concept of Consciousness

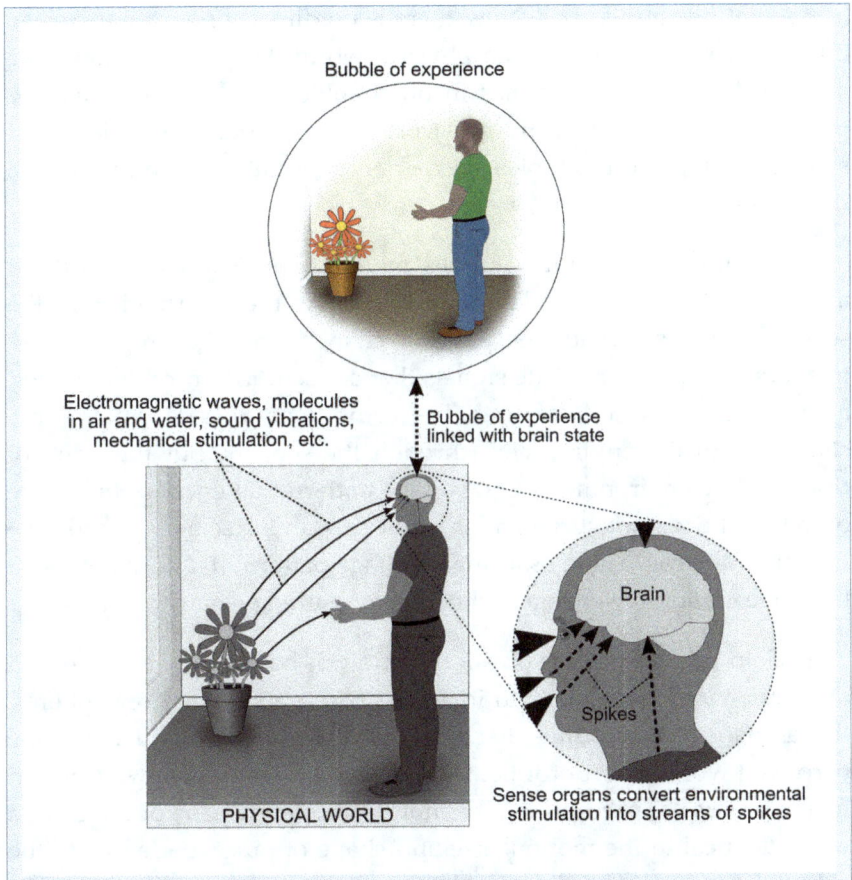

Figure 2.5. The relationship between a bubble of experience and a brain. I have assumed that the brain is the only part of the body that is linked to a bubble of experience. Signals from the world interact with the sense organs, which send streams of spikes down the nerves to the brain. The resulting brain state is linked with a bubble of experience. Image © David Gamez, CC BY 4.0.

an online bubble of experience directly correspond to primary qualities in the world. The pungent bear-smell did not exist in the physical world; the size, shape and movement of the bear in my bubble of experience matched or *resembled* the size, shape and movement of the physical bear.

Modern science has interposed the brain between bubbles of experience and objects in the physical world. Our experiences of size, shape and movement are now thought to be linked to firing patterns in populations of neurons. Now there is no direct connection between bubbles of experience and the physical world. Why, then, should we assume that primary qualities in our bubbles of experience *resemble* primary qualities in the physical world? Why should we believe that space and time in our bubbles of experience resemble space and time in the physical world?[14]

A computer is driving a car. Its memory consists of voltages that are updated by cameras, lasers and GPS. As the information in the sensors changes the voltage patterns change, and the program uses this information to calculate signals that are sent to control the brakes, accelerator, gears and steering. The computer's voltage patterns are connected to the environment through the sensors, but they do not resemble the environment. The voltage pattern that encodes the shape of the road does not curve and it is not the same size as the road. The motion of the car is held as a single voltage pattern that does not move like the car and only changes when the measured velocity changes.

Back in Locke's day the physical world was believed to be composed of atoms, which were easy to imagine as tiny bouncing grey spheres. It was natural to assume that the physical world was just like the perceived world, except for the secondary qualities, which were added by the process of perception. The motion, size and shape of the objects were identical to the motion, size and shape of our experiences of the objects—we were indeed seeing the things themselves.

Today the physical world has become unimaginable. We cannot imagine what a wave-particle or a ten-dimensional superstring is *like*. We have lost all reasons for believing in resemblance between our bubbles of experience and the physical world. We have no grounds for attributing either the primary or the secondary qualities of our bubbles of experience to the invisible world described by modern physics.[15]

We cannot prove that a physical bear is *not* identical to the appearance of a bear in a bubble of experience. And we have little reason to believe that a physical bear *does* resemble a bear in a bubble of experience. *We just don't know and cannot know*. We cannot reach beyond our senses to see the physical world as it is in itself. We have to suspend judgment about what the physical world is *really like*.[16]

2. The Emergence of the Concept of Consciousness

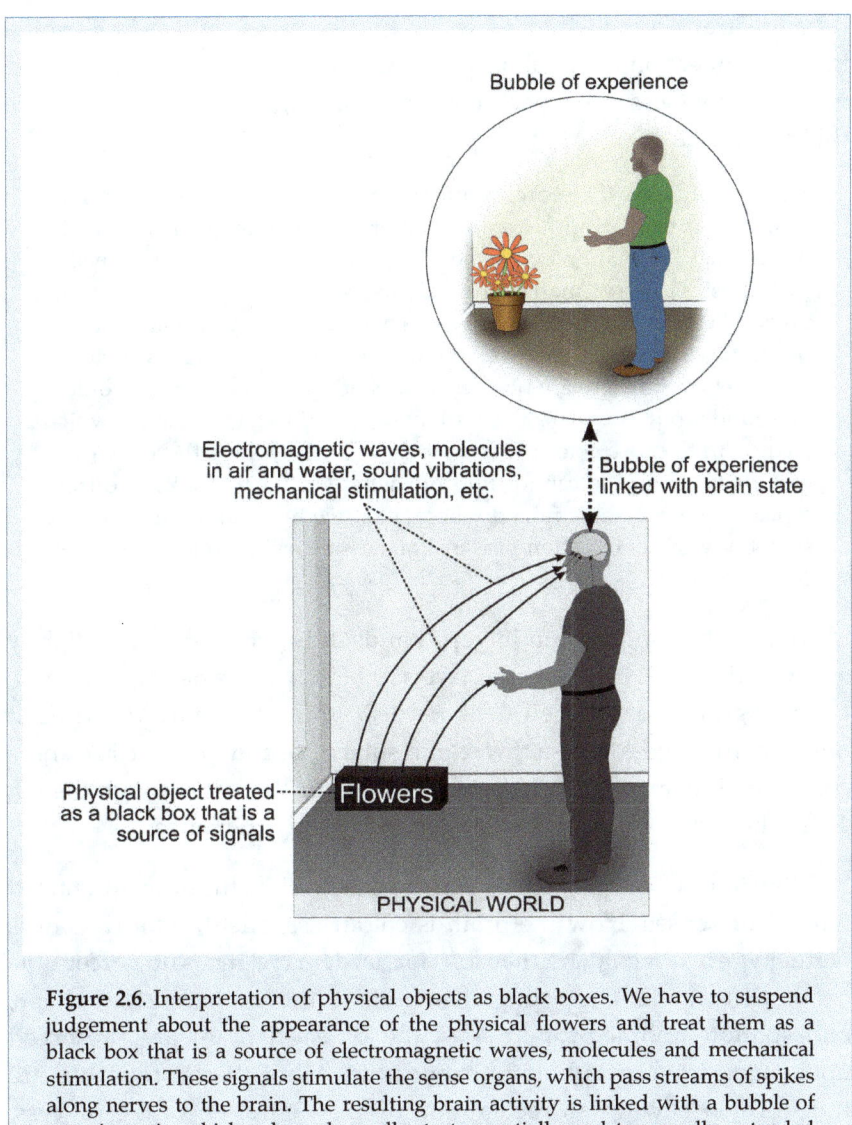

Figure 2.6. Interpretation of physical objects as black boxes. We have to suspend judgement about the appearance of the physical flowers and treat them as a black box that is a source of electromagnetic waves, molecules and mechanical stimulation. These signals stimulate the sense organs, which pass streams of spikes along nerves to the brain. The resulting brain activity is linked with a bubble of experience in which coloured, smelly, tasty, spatially and temporally extended flowers appear. Image © David Gamez, CC BY 4.0.[17]

As far as *we* are concerned physical objects are black boxes that interact with each other in accordance with the laws of physics. They are also sources of signals that enter our senses and are processed into spiking patterns that are sent along nerves to our physical brains, where they are transformed into more spiking patterns, which have some kind of connection with bubbles of experience that contain the coloured warm smelly faces of the people we love (see Figure 2.6). Russell makes this point well:

> Modern physics, therefore, reduces matter to a set of events which proceed outward from a centre. If there is something further in the centre itself, we cannot know about it, and it is irrelevant to physics. [...] Physics is mathematical, not because we know so much about the physical world, but because we know so little: it is only its mathematical properties that we can discover. For the rest, our knowledge is negative. In places where there are no eyes or ears or brains there are no colours or sounds, but there are events having certain characteristics which lead them to cause colours and sounds in places where there are eyes, ears and brains. We cannot find out what the world looks like from a place where there is nobody, because if we go to look there will be somebody there; the attempt is as hopeless as trying to jump on one's own shadow.[18]

You are holding a brain in your hands: it is soft, warm and slightly sticky with blood. You lick it. It tastes of blood. You smell it—a humid, fresh, slightly meaty smell. It is reddish grey in colour. You drop it onto a marble worktop—a thwacking splat of sound. It has a size and a convoluted texture. It moves when slapped or thrown through the air. This is how the brain appears in your bubble of experience.

Now remove the properties that appeared when the brain interacted with your senses. Now the brain is colourless, silent, odourless; it is neither warm nor cold, neither soft nor hard; in fact it has no perceptible properties at all. Drop the illusion that the motion, size, shape, and spatiotemporal properties of the physical brain are preserved unchanged in your bubble of experience. All of these properties are transformed beyond all recognition by the neural encoding process. The physical brain vanishes: it can no longer appear as it is in itself in your bubble of experience. As far as you are concerned, physical brains are black boxes, just like every other object in the physical world. This is illustrated in Figure 2.7.

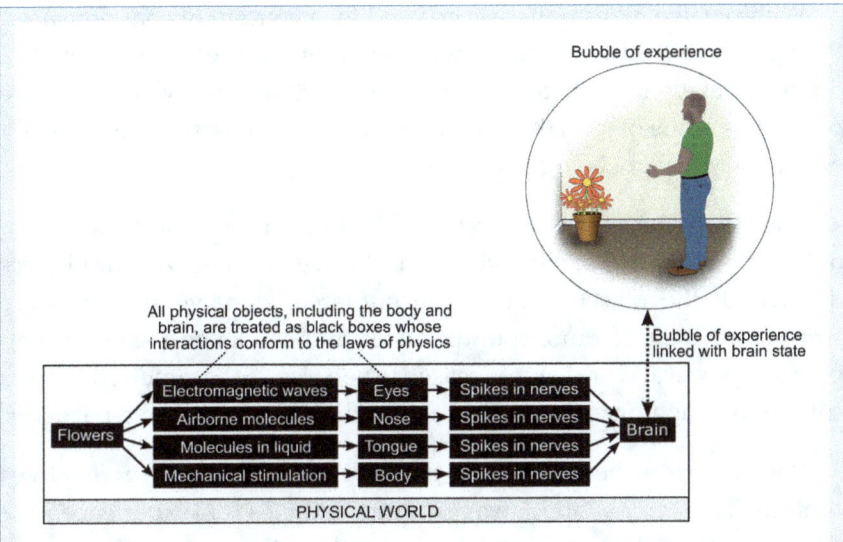

Figure 2.7. The relationship between a bubble of experience and an invisible physical brain. As far as we are concerned all objects in the physical world, including our bodies and brains, are black boxes. The arrows show the interactions between these objects in accordance with the laws of physics. Objects in the environment are sources of signals that lead to brain activity that has some kind of connection with a bubble of experience. Image © David Gamez, CC BY 4.0.

2.5 The Emergence of the Concept of Consciousness

I look to the right and see a dirty grey wall. I look to the left and see a black lamppost with scratches and flaking paint. Ahead of me a decrepit old man hobbles along a derelict street. I am awake, not dead, but consciousness is not present anywhere—there is no consciousness in the man's stained trousers, no consciousness in the dirt, no consciousness in the smell of dog piss. Nor does consciousness appear when I turn my attention to my ulcerating stomach and painful feet. Consciousness is completely absent from my bubble of experience as I view the street. My reports are not driven by a *thing* or *property* called consciousness. I can describe everything without mentioning consciousness once.

Within naive realism there is no need for a concept of consciousness. People perceive different things with different levels of clarity and have different levels of wakefulness. Primary and secondary qualities are properties of the objects themselves, which they possess independently of whether they are being perceived.

The scientific revolution revived atomic theories and led to an unimaginable world of superstrings and wave-particles. As our physics developed, the objects that we encountered in naive realism were stripped of their colours, sounds, tastes and smells and sank into an invisible physical world. A tree ceased to be a tree—it became a colourless collection of jigging atoms, a probability distribution of wave-particles.

Physical trees became black box sources of signals; green trees continued to creak and sway in our bubbles of experience. We attempted to explain them away, and yet there they were in front of us with properties that could not be neatly shoehorned into the world of physics. We had to find a way of grouping, describing and explaining the colourful, smelly, noisy properties that were originally attributed to objects in naive realism.

We solved this problem by inventing the modern concept of consciousness. 'Consciousness' became a name for bubbles of experience, which were reinterpreted in relation to an invisible physical world. This is formally stated as follows:[19]

> **D1.** *Consciousness* is another name for *bubbles of experience*. A state of a consciousness is a state of a bubble of experience.[20] Consciousness includes all of the properties that were removed from the physical world as scientists developed our modern invisible explanations.

Initially the modern concept of consciousness emerged in response to the renaissance of atomism. In the seventeenth century the physical world was believed to only have primary qualities—secondary qualities were excluded from this world and developed a separate existence of their own that demanded an explanation. The solution was to package up secondary qualities with the concepts of mind, thinking substance and consciousness. This interpretation of consciousness is nicely summarized by Galileo:

Now I say that whenever I conceive any material or corporeal substance, I immediately feel the need to think of it as bounded, and as having this or that shape; as being large or small in relation to other things, and in some specific place at any given time; as being in motion or at rest; as touching or not touching some other body; and as being one in number, or few, or many. From these conditions I cannot separate such a substance by any stretch of my imagination. But that it must be white or red, bitter or sweet, noisy or silent, and of sweet or foul odour, my mind does not feel compelled to bring in as necessary accompaniments. Without the senses as our guides, reason or imagination unaided would probably never arrive at qualities like these. Hence I think that tastes, odours, colors, and so on are no more than mere names so far as the object in which we place them is concerned, and that they reside only in the consciousness. Hence if the living creature were removed, all these qualities would be wiped away and annihilated. But since we have imposed on them special names, distinct from those of the other and real qualities mentioned previously, we wish to believe that they really exist as different from those.[21]

The twentieth century developed the concept of consciousness to its logical conclusion. Our theories about the physical world became mathematical and abstract—they make beautiful predictions, but they are no longer based on the everyday properties and objects that we encounter in our bubbles of experience. The twentieth century also developed theories about how bubbles of experience are linked to the brain. This eliminated our reasons for believing that primary qualities in our bubbles of experience resemble primary qualities in the physical world. While physics is perceived to be the true or ultimate reality, we continue to be immersed in bubbles of experience in our daily lives: our need to express and address this issue led to the modern concept of consciousness. This trajectory from naive realism to twentieth century science and consciousness is illustrated in Figure 2.8.

The *contents* of a person's consciousness are the objects and properties in their bubble of experience. When a burning bush is in my bubble of experience, the colour, smell, taste, heat and sound of the burning bush are the contents of my consciousness. When I say 'I am conscious of hissing sap and orange flames,' I am stating that hissing sap and orange flames are in my bubble of experience.

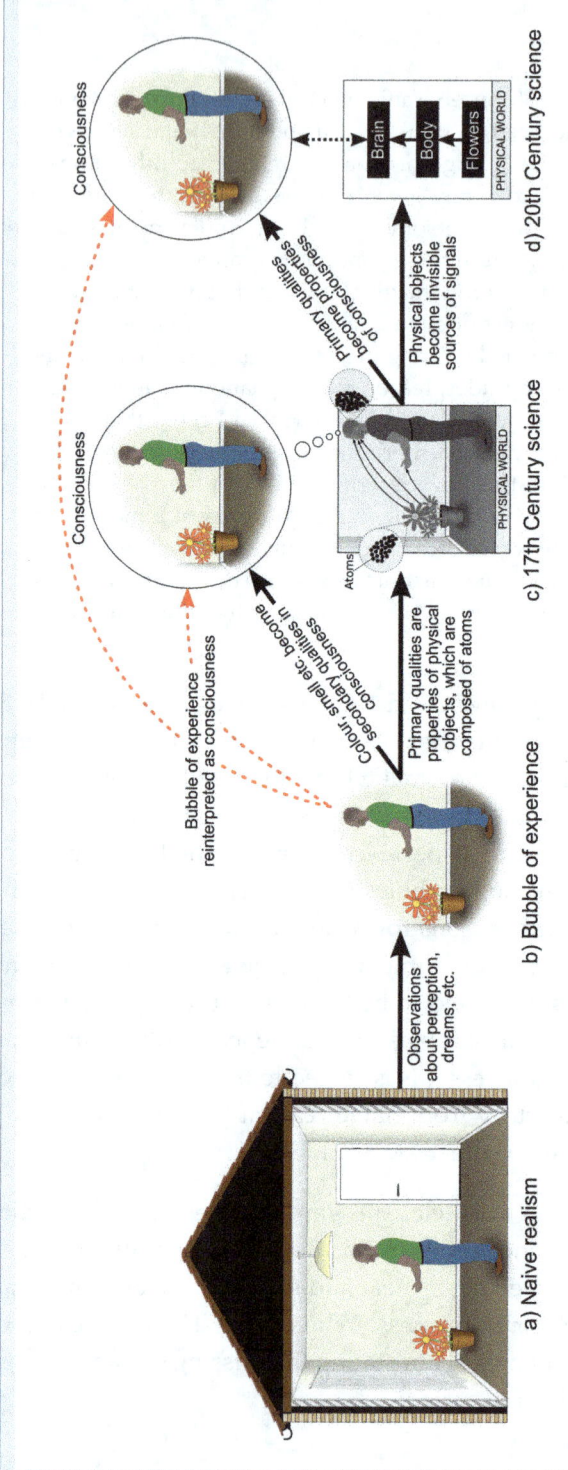

Figure 2.8. The emergence of the concept of consciousness. a) Naive realism. Objects have the properties that they are perceived to have and continue to have these properties when they are not being perceived. b) Naive realism is supplemented with a theory of perception, which I described using the idea of a bubble of perception. This was reinterpreted as a bubble of experience to handle dreams, hallucinations, etc. c) The revival of atomism in the seventeenth century led to a distinction between primary qualities, which are properties of physical objects, and secondary qualities that arise when the physical world interacts with the senses. The concept of consciousness was invented to accommodate the secondary qualities that were excluded from the physical world. d) Twentieth century science eliminates all resemblance between bubbles of experience and the physical world. Everything in our bubbles of experience is interpreted as consciousness. Image © David Gamez, CC BY 4.0.

Many consciousness experiments are based on the idea that a person has a particular *level* of consciousness. A person's overall level of consciousness can be defined as the average intensity of the contents of their bubble of experience.[22] A high intensity, vivid, stable bubble of experience with high resolution is consciousness at a high level. A bubble of experience that contains a few faint and unstable objects is consciousness at a low level. We say that Zampano is conscious when his physical brain is associated with a bubble of experience that has non-zero intensity.[23] We say that Zampano is unconscious when his physical brain is not associated with a bubble of experience.

The distinction between online and offline bubbles of experience can be expressed in terms of online and offline conscious contents:[24]

- *Online* conscious contents are linked to states of the environment and are updated in response to changes in the environment. The environment is functionally connected to online conscious contents.[25]
- *Offline* conscious contents are independent of the environment. There is no functional connection between the current environment and offline conscious contents.

Consciousness can contain a mixture of online and offline contents. When I worship at the tombs of my ancestors the shadowy form of my grandfather rises from his grave, winks and raises his hat. My grandfather and his hat are offline conscious contents; the tombs and surrounding graveyard are online conscious contents.[26]

A suggestive piece of evidence for a link between the rise of science and the emergence of the concept of consciousness is Wilkes' observation that there was no word for consciousness in the English language prior to the seventeenth century or in ancient Greek or Chinese:

> Two intriguing facts. First, the terms 'mind' and 'conscious(ness)' are notoriously difficult to translate into some other languages. Second, in English (and other European languages) one of these terms—'conscious' and its cognates—is in its present range of senses scarcely three centuries old. [...] In ancient Greek there is nothing corresponding to either 'mind' or 'consciousness' [...] In Chinese, there are considerable problems in capturing 'conscious(ness)'. And in Croatian, 'mind' poses interesting difficulties.[27]

There are contexts in which our modern English word for consciousness can be translated into ancient Greek or Chinese—for example, by 'psyche', 'Sophia', 'nous', 'metanoia' or 'aesthesis' in ancient Greek, or by 'yìshì' in Chinese. However, Wilkes claims that there is no generally adequate translation that captures our current use of 'consciousness'.

According to Wilkes, this linguistic data shows that the modern concept of consciousness covers a number of disparate phenomena:

- Whether someone is awake or asleep.
- Body sensations, such as itches and pains.
- Sensory experience—for example, colours, tastes and smells.
- Ascription of propositional attitudes, such as deliberating, pondering, desiring and believing.

This leads Wilkes to conclude that consciousness is unlikely to be a natural kind or something that we can study scientifically:

> Essentially, I am trying to say two distinguishable things. First, that in all the contexts in which it tends to be deployed, the term 'conscious' and its cognates are, for *scientific* purposes, both unhelpful and unnecessary. The assorted domains of research, so crudely *indicated* by the ordinary language term, can and should be carved up into taxonomies that cross-classify those which emphasis on 'consciousness' would suggest. Second, that we have little if any reason to suppose that these various domains have anything interesting in common: that is, consciousness will not just be a (cluster) natural kind.[28]

However, Wilkes' observations about 'consciousness' can be interpreted to support the idea that consciousness is a modern name for a bubble of experience. Bubbles of experience are natural kinds that are common to all people speaking all languages. We all see red objects, feel heat, smell flowers and taste meat. However, scientific theories about an invisible physical world are a recent product of a great deal of conceptual, technological and experimental effort. Earlier societies lacked our interpretation of physical reality, so it is not surprising that our modern concept of consciousness is absent from ancient Greek, Chinese and the English language prior to the seventeenth century. Bubbles of experience were once understood in relation to an invisible world of gods and spirits. Once we started to believe in a physical world of atoms and forces, the colourful conscious world (that is manifestly *not* composed of atoms and forces) emerged as a separate area of inquiry.

One potential problem for this interpretation of consciousness is that it was not developed by the ancient atomists. Since they believed that colours, sounds and smells are not properties of atoms, it would have been natural to place them in a second substance, such as mind or consciousness. So why was consciousness (or a similar concept) absent from ancient Greek, but developed by atomists in the seventeenth century?

This problem would be resolved if the ancient atomists did invent a word for consciousness that did not enter common usage and was lost, or if they expressed the concept in a more indirect way. Although very little material is left from the ancient Greek atomists, it might be possible to find traces of a concept of consciousness in their work. A second possibility is that the ancient atomists might have believed that secondary qualities could be reduced to primary qualities. Plenty of people today believe that consciousness can be reduced to the physical world, so it would not be surprising if the ancient atomists had a similar view.[29] It is also possible that the consequences of ancient atomism were not fully worked out. At the time atomism was one of a large number of highly speculative theories about the world and it is conceivable that the small number of people who believed in atomism did not have the time or resources to develop it fully. Today our theories about the physical world are subject to wide agreement, which has led to a general need for a concept of consciousness to contain the properties that have been excluded from the physical world.

2.6 Summary

Most of our lives are spent in a state of naive realism in which we attribute colours, sounds and smells to objects in our environment. I developed the concept of a bubble of experience to describe how we only perceive part of the world at one time, and to accommodate observations about dreams, imagination and hallucination.

The physical world is an invisible source of signals that interact with our sense organs to produce patterns in our brains that are somehow connected with our bubbles of experience. There is unlikely to be any resemblance between the contents of our bubbles of experience and the physical world.

When science eliminated sensory properties from the physical world it was necessary to find a way of grouping, describing and explaining the colours, sounds and smells that we continue to encounter in daily life. We solved this problem by inventing the modern concept of consciousness. 'Consciousness' is another name for our bubbles of experience, which contain the sensory properties that science removed from the physical world.

3. The Philosophy and Science of Consciousness

3.1 Understanding Consciousness

How should I understand and explain my wife? I can describe the changes in her hair and skin colour as she moves about in the light. I can objectify her body or relate to her as the Other—she faces me, crushes me, makes me feel *guilty*. I can use Eros and Thanatos to analyze her psychology, or interpret her mind as an intricate neural mechanism. I can provide an evolutionary explanation of her features (noting a trace of *Homo heidelbergensis* in her face).

How should we understand and explain consciousness?

Phenomenologists bracket off the physical world and describe the structure of consciousness from a first-person perspective.[1] The starting point for phenomenology is our immersion in conscious experience; the end point is a description of consciousness that sets aside scientific theories about the physical world.

Science has triumphed in the last three hundred years. We send people to the moon and grow babies in test tubes. Many people believe that science provides a complete description of the world. The predictive success of science exerts a commanding weight: we believe in the physical world; we are convinced by our scientific explanations and cannot ignore them. We need phenomenological descriptions of consciousness, but we cannot bracket out scientific theories. Phenomenology is not enough.

Many people explain consciousness by reducing it to features of the physical world.[2] This is not convincing. It is far from obvious that

© David Gamez, CC BY 4.0   http://dx.doi.org/10.11647/OBP.0107.03

colours, sounds and smells can be reduced (with a wave of the hand) to wave-particles, superstrings or fields. Colour appears in consciousness when colourless electromagnetic waves interact with the physical sense organs and change the brain's activity. But brain activity is silent and without smell. Colour and smell are linked to brain activity; they are not identical to colourless odourless brain activity.

Bubbles of experience and the physical world are both important phenomena. It would be premature to bracket either out or to claim that one can be reduced to the other. *Both* have to be taken *seriously*. We have detailed descriptions of consciousness and a good understanding of the physical world—the key gap in our knowledge is the relationship between them.

3.2 The Limits of Imagination

Imagination is an offline bubble of experience.[3] We can change the contents of this bubble of experience without changing our physical environment. We fill it with novel things—recombine the elements of our experience in new ways. As I sit at my desk I imagine that I have lost my limbs and lie naked on a rocky plain beneath a burning sky. A raven is feeding on my liver.

We use our imagination to solve scientific problems. On separate unrelated occasions mutating DNA, tumours, cigarette smoke and unregulated cell growth appear in my bubble of experience. I use my empirical knowledge about the links between these phenomena to visualize the correct sequence from smoking to the appearance of a tumour. I do not need direct knowledge of the physical world to do this. It is enough that the appearance of cigarette smoke in my bubble of experience corresponds to the presence of smoke in the physical world, that the appearance of mutating DNA in my bubble of experience corresponds to the presence of mutating DNA in the physical world, and so on (see Figure 3.1). We have an intuitively satisfying explanation of the relationship between phenomena when we can visualize the intermediate steps between them.

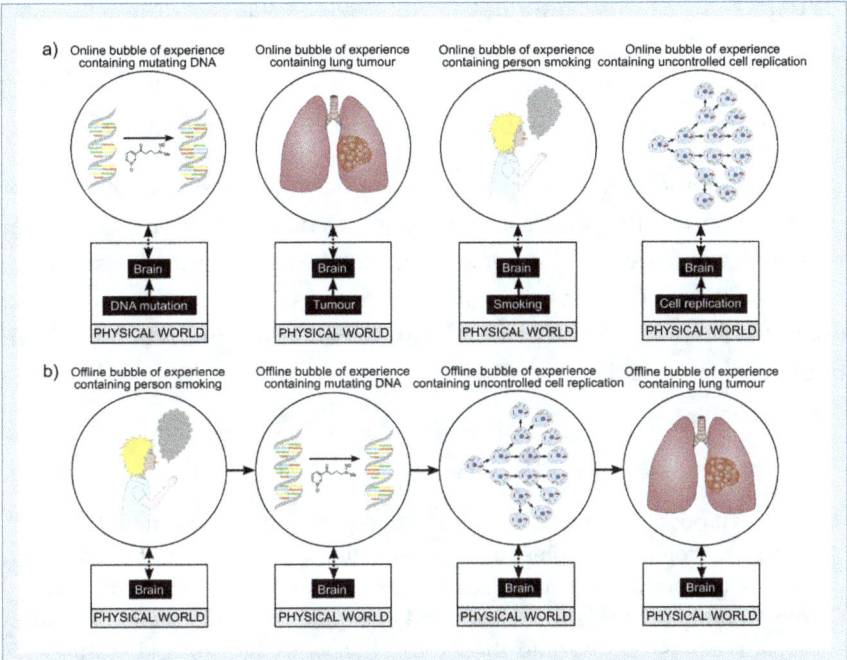

Figure 3.1. The use of imagination to solve a scientific problem. a) On separate occasions mutating DNA, tumours, cigarette smoke and unregulated cell growth appear in my bubble of experience. I observe sequential relationships between pairs of these phenomena, but never the whole story. b) My empirical knowledge about the connections between these phenomena enables me to imagine the correct sequence of steps from smoking to the appearance of a tumour in the lungs. Image © David Gamez, CC BY 4.0.

Online and offline bubbles of experience have the same properties—colour, smell, taste, sound and body sensations. These are typically less intense in offline bubbles of experience and some are not present at all. Objects can vary in wild ways in an offline bubble of experience, but we cannot imagine new properties. We cannot imagine properties that we have not encountered in an online bubble of experience.

Imagination is an inductive engine that reassembles previous experiences. We can imagine pigs playing the piano because we have seen pigs and pianos before. If we had never seen pianos and pigs, then it is unlikely that piano-playing pigs would enter our imagination.[4]

We have a limited ability to wilfully transform our offline bubbles of experience. It is difficult to imagine radically different forms of consciousness. It is difficult or impossible to wilfully morph our bubbles of experience into a bat's bubble of experience (assuming it has one).[5]

We cannot imagine things that cannot become conscious. We cannot imagine an invisible physical world that has none of the properties that we encounter in our bubbles of experience. We can imagine large brains, small brains, blue brains, green brains, brains made of cheese, and so on. But the physical brain cannot be imagined as it is in itself, outside all bubbles of experience.[6]

3.3 A Failure of Imagination

How is it possible for conscious states to depend upon brain states? How can technicolour phenomenology arise from soggy grey matter? What makes the bodily organ we call the brain so radically different from other bodily organs, say the kidneys—the body parts without a trace of consciousness? How could the aggregation of millions of individually insentient neurons generate subjective awareness? We know that brains are the *de facto* causal basis of consciousness, but we have, it seems, no understanding whatever of how this can be so. It strikes us as miraculous, eerie, even faintly comic. Somehow, we feel, the water of the physical brain is turned into the wine of consciousness, but we draw a total blank on the nature of this conversion. Neural transmissions just seem like the wrong kind of materials with which to bring consciousness into the world, but it appears that in some way they perform this mysterious feat.

Colin McGinn, Can We Solve the Mind-Body Problem?[7]

In 1989 the philosopher Colin McGinn asked the following question "How can technicolor phenomenology arise from soggy gray matter?" [...] Since then many authors in the field of consciousness research have quoted this question over and over, like a slogan that in a nutshell conveys a deep and important theoretical problem. It seems that almost none of them discovered the subtle trap inherent in this question. The brain is not gray. The brain is colorless.

Thomas Metzinger,
Consciousness Research at the End of the Twentieth Century[8]

With furrowed brow, buried deep in his leather-bound armchair, the Philosopher seeks to untangle the mysteries of consciousness. One of the greatest and hardest problems of his time, or so he has read. He imagines the physical world, a grey lifeless sort of place. He imagines the brain, a squishy grey thing floating dismally in the air above his dirty glass coffee table. He imagines the colour RED. Bracing himself for the challenge, he attempts to solve the Hard Problem of Consciousness. Many have failed before him, but his obvious brilliance and intellectual arrogance will surely enable him to triumph. Perspiration breaks out on his brow. 'Gee this is a tough one.' He struggles to imagine how activity in the grey brain could cause or *be* the colour red. Or how the colour red could be reduced to activity in the grey brain. He changes position, scratches his crotch, picks his nose. He is sure that he can solve it, write that brilliant paper, receive the rapturous admiration of his peers. 'Just a bit more time; perhaps a coffee will help.'

The relationship between consciousness and the physical world is often addressed in a thought experiment in which we bring to mind an image of the brain (coloured grey) and an image of something conscious (the colour red), and try to imagine how they are related. This is often described as a *hard* problem of consciousness because it is impossible for us to imagine how neural activity could cause or *be* the colour red.[9]

The physical brain has none of the properties that are present in consciousness. It is not grey; it is not soggy—it is invisible and cannot be imagined by us. So thought experiments and imagination cannot be used to study the relationship between invisible physical brains and conscious experiences. They can *only* be used to study the relationship between our conscious experiences of brains and other conscious experiences. This is illustrated in Figure 3.2.

3.4 Regularities in Conscious Experiences

We can observe *regularities* in our conscious experiences. Suppose I am connected to a device that shows the state of my brain on a screen. When an ice cube is placed in my left hand I observe brain pattern p_1 on the screen. When I have memorized p_1 I can make an imaginative transition from p_1 to a cold sensation: whenever I think about p_1 I imagine an ice cube. This is illustrated in Figure 3.3.[10]

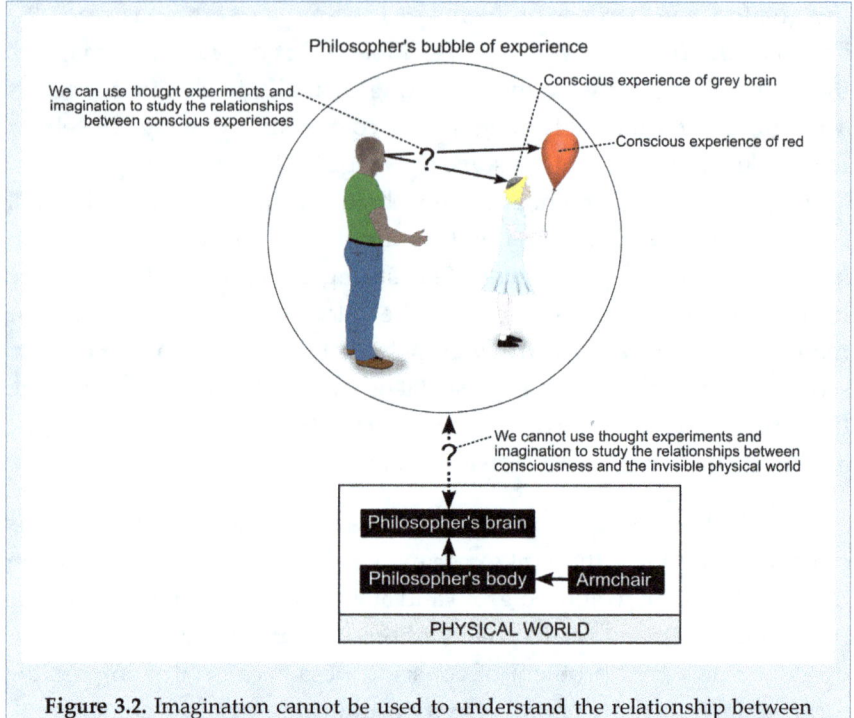

Figure 3.2. Imagination cannot be used to understand the relationship between consciousness and the invisible physical world. A philosopher imagines a child with her brain exposed holding a red balloon. He wants to understand the relationship between the colour red and the invisible physical brain, but he can only imagine the relationship between the colour red and his conscious experience of a grey brain. Image © David Gamez, CC BY 4.0.

We find it hard to imagine how our conscious experiences of brains are related to other conscious experiences because we have had little exposure to this relationship.[11] Imagination is an inductive engine—it needs to be exposed to an association between A and B before it can imagine a transition from A to B. As technology develops we will have more conscious experiences of brain activity, which will enable us to develop intuitive links between our conscious experiences of brain patterns and other conscious experiences. This is nicely illustrated by Rorty's example of the Antipodeans:[12]

> In most respects, then, the language, life, technology, and philosophy of this race were much like ours. But there was one important difference. Neurology and biochemistry had been the first disciplines in which

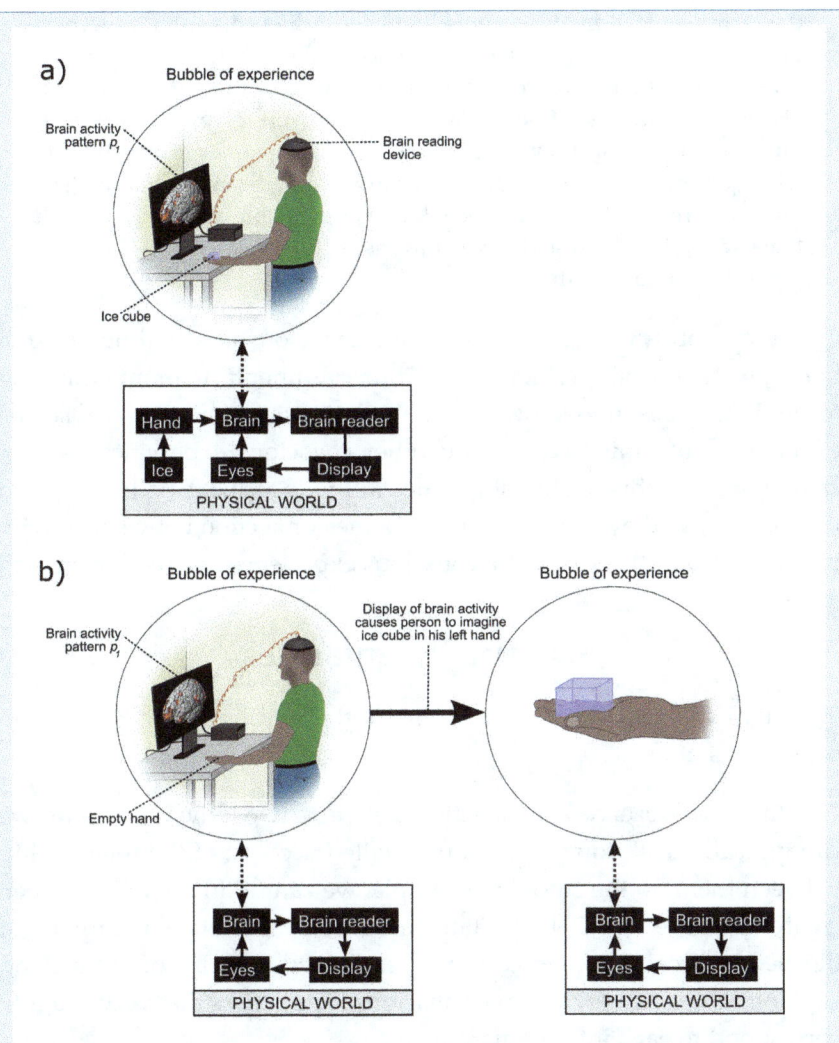

Figure 3.3. Learnt association between consciously experienced brain activity and the sensation of an ice cube. a) The subject wears a device that displays his brain activity on a screen. When an ice cube is placed in his left hand he observes and memorizes the brain pattern, p_1, that appears on the screen. b) At a later time the subject views p_1 when he is not holding an ice cube. He makes an imaginative transition from his conscious experience of p_1 to a conscious experience of an ice cube. Image © David Gamez, CC BY 4.0.

the technological breakthroughs had been achieved, and a large part of the conversation of these people concerned the state of their nerves. When their infants veered towards hot stoves, mothers cried out, "He'll stimulate his c-fibers." When people were given clever visual illusions to look at, they said, "How odd! It makes neuronic bundle G-14 quiver, but when I look at it from the side I can see that its [sic] not a red rectangle at all." Their knowledge of physiology was such that each well formed sentence could easily be correlated with a readily identifiable neural state.[13]

We can observe regularities in our consciousness without raising hard philosophical problems. But our rudimentary brain scanning technology does not show us the relationship between conscious experiences of brain activity and other conscious experiences, so we cannot imagine this relationship. The Antipodeans have better access to their brains, so they can easily imagine the connection between activity in neuronic bundle G-14 and a conscious experience of a red rectangle.[14]

3.5 Brute Regularities

We might still ask *how* or *why* activity in the physical brain is linked to conscious states.

When we observe a connection between two physical events we can typically drill down to a more detailed account of the relationship between them. In the smoking example, we can fill in the link between cigarette smoke and DNA mutations with a more detailed story about biochemical reactions, which can be explained in terms of atomic and subatomic events. We can have conscious experiences that very roughly correspond to each intermediate stage.

In the case of consciousness we are looking for a relationship between something that is physical and properties that are not attributed to the physical world (colour, smell, etc.). However much data we gain about consciousness and the brain there is bound to be a gap in the imaginative story. We can learn everything there is to know about links between conscious experiences of brain activity and other conscious experiences, but we will still reach a point at which we simply have to accept that a particular brain state is connected with a particular conscious state. A physical brain activity pattern could simply be associated with red—this might be a *brute regularity* that cannot be broken down and analyzed any further.

Brute regularities exist in the other sciences. At a certain point we simply have to accept that the physical world works in a way that we can describe but cannot explain. For example, the behaviour of superstrings or elementary wave-particles can be described but not explained—it is the starting point for higher level physical explanations.[15] Scientists cannot say why physical brute regularities exist. They are simply how the universe works.

Physics gives a detailed hierarchical description of the relationships between physical things, ranging from the interactions between elementary wave-particles up to the behaviour of planets and galaxies. Brute regularities lie at the bottom of this hierarchy—at the level of superstrings and elementary wave-particles.[16]

Our understanding of the relationship between consciousness and the physical world is at an early stage of development. We have no idea what the brute regularities of consciousness science will be. They could be simple relationships between novel physical properties and atoms of conscious experience, or they could be more complex regularities linking distributed brain activity patterns to complex conscious experiences.

Some people think that brute regularities are hard problems. But genuine brute regularities are not problems at all. They are basic facts about the way the universe works. Other phenomena pose problems that can be solved in terms of brute regularities. The brute regularities themselves can only be described—they cannot be understood or explained.

3.6 Summary

Idealists reject the physical world; phenomenologists suspend judgement about it; physicalists claim that consciousness *is* the physical world. None of these positions is convincing. Consciousness and the physical world *both* have to be taken seriously as real phenomena. We can study the relationship between them and suspend judgement about the metaphysical debates.

We cannot imagine the invisible physical world. So thought experiments and imagination cannot be used to study the relationship between invisible physical brains and conscious experiences. They can *only* be used to study the relationship between our conscious experiences of brains and other conscious experiences. As brain-scanning technology improves we will find it easier to make imaginative transitions between conscious experiences of brain states and other conscious experiences.

At some point the science of consciousness will encounter brute regularities in the relationship between consciousness and the physical world that can be described, but not explained. Brute regularities exist in the other sciences. We have no idea what the brute regularities of consciousness science will be.

The rest of this book suggests how a systematic science of consciousness can be developed. We can measure consciousness, measure the physical world and develop mathematical theories of the relationship between these measurements. Scientists measure consciousness using first-person reports, which raises traditional philosophical problems of zombies, solipsism, colour inversion and inaccessible consciousness. These problems cannot be solved, but they can be neutralized using the framework of definitions and assumptions that is set out in the next chapter.

4. The Measurement of Consciousness

> Consciousness just is not the sort of thing that can be measured directly. What, then, do we do without a consciousness meter? How can the search go forward? How does all this experimental research proceed?
>
> I think the answer is this: we get there with principles of *interpretation*, by which we interpret physical systems to judge the presence of consciousness. We might call these *preexperimental bridging principles*. They are the criteria that we bring to bear in looking at systems to say (1) whether or not they are conscious now, and (2) which information they are conscious of, and which they are not.
>
> David Chalmers,
> On the Search for the Neural Correlates of Consciousness[1]

4.1 First-Person Reports about Consciousness (C-Reports)

I am standing with my friend Olaf in a field of poppies. 'Look Olaf,' I say, 'the poppies are red, the sky is blue and the leaves are green.' 'By the blood of Grendel,' he replies, 'I can hear the sound of a bird singing and feel a sensation of warmth in my left foot.'

In earlier times my chat with Olaf would have been interpreted as a conversation about the world. Over the last three hundred years science has sucked colour, sound and warmth out of the world and reinterpreted them as consciousness. Statements like 'The poppies are red' or 'There is a rusty helmet on the ground' have become descriptions of consciousness.[2]

I am *certain* that I can speak about my consciousness. I cannot doubt that 'The poppies are red' is a true statement about my bubble of experience. I would be more willing to jettison the entire edifice of science, than abandon my belief that I can describe my consciousness.[3]

I can speak about my consciousness. I can describe my consciousness by pushing buttons and pulling levers. I can reply to questions about my consciousness by putting my brain into different states in a fMRI scanner.[4]

Olaf is alert. He can flexibly respond to novel situations. He can inwardly execute a sequence of problem-solving steps. He can execute a delayed reaction to a stimulus and respond to verbal commands.[5] He is willing to bet a large amount of money that there is a rusty helmet in the field of poppies.[6] These behaviours can be used to make reliable inferences about the contents and level of Olaf's consciousness, even when he is not explicitly reporting his consciousness.[7]

I punch Olaf in the face. He falls to the ground and lies still. His stillness and lack of response are external signs that his brain is not associated with a bubble of experience, that his level of consciousness is zero.

When Olaf regains consciousness he exhibits groggy behaviour. I interpret this as a sign that he has a low level of consciousness. He is never quite the same again and often behaves in a similar way to a patient described by Damasio:

> Suddenly the man stopped, in midsentence, and his face lost animation; his mouth froze, still open, and his eyes became vacuously fixed on some point on the wall behind me. For a few seconds he remained motionless. I spoke his name but there was no reply. Then he began to move a little, he smacked his lips, his eyes shifted to the table between us, he seemed to see a cup of coffee and a small metal vase of flowers; he must have because he picked up the cup and drank from it. I spoke to him again and again he did not reply. He touched the vase. I asked him what was going on and he did not reply, his face had no expression. [...] Now he turned around and walked slowly to the door. I got up and called him again. He stopped, he looked at me, and some expression returned to his face—he looked perplexed. I called him again and he said, "What?"[8]

When Olaf is in this state he is not capable of executing a sequence of problem-solving steps. He does not flexibly respond to novel situations.

He cannot execute a delayed reaction to a stimulus. We interpret his behaviour as a sign that he has zero consciousness, that he is not immersed in a bubble of experience.

Any behaviour that can be interpreted as a measurement of the level and/or contents of a person's consciousness will be referred to as a *c-report*:

D2. A *c-report* is a physical behaviour that is interpreted as a report about a person's consciousness.

A c-report is a *measurement of consciousness*. This measurement is indirect—Olaf's bubble of experience does not appear in my bubble of experience.

Indirect measurements are standard scientific practice. When I measure the path of a particle, the particle does not directly appear in my bubble of experience. I have to create an experimental situation in which the particle creates a visible trace, such as a track of bubbles in a chamber. Theories about the physical world link the bubble track to the path of the invisible particle.

4.2 Reports about Non-Conscious Mental Content (NC-Reports)

Olaf thinks a lot about his sweetheart Olga. As he crosses the field of poppies he is thinking about her corn-blond plaits, her inviting smile, her chequered billowing skirt, her strong smooth thighs. He is not aware of the stones in his boots, the white crosses in the field or the hot sun on his face. None of these are in his bubble of experience, although he could bring them into his bubble of experience if he stopped thinking about Olga's thighs and focused on his body and surroundings.

As Olaf walks and thinks about Olga, the sensory data from the field of poppies is used by his brain to generate control signals that are sent to his muscles. This sensory data does not appear in his bubble of experience. It is *unconscious* or *non-conscious* information.

I present a picture of Olga to Olaf's right eye and a picture of Olaf's ex-wife Ingrid to his left eye. He experiences a phenomenon called binocular rivalry in which Olga's picture is perceived for a few seconds while Ingrid's is non-conscious, and then Ingrid's picture becomes conscious and Olga's non-conscious. When Ingrid's picture is non-conscious it is still being processed by Olaf's brain, which responds to the shape of her sharp tongue in her hard mouth.[9]

I show Ingrid's picture to Olaf for 30 ms in the middle of a sequence of scrambled images. Under these conditions Ingrid's picture does not enter Olaf's bubble of experience, but it does cause Olaf to complete word fragments with Ingrid-related words and alters the conductivity of his skin.[10] When I ask Olaf to guess which picture was shown he picks Ingrid's picture more often than chance.[11]

All of these behaviours can be used to identify mental contents that are being processed non-consciously. They are *nc-reports*:

D3. A *nc-report* is a physical behaviour that is interpreted as a report about non-conscious mental content.

4.3 Platinum Standard Systems

C-reports about consciousness can be found everywhere. The sigh of waves can be interpreted as a c-report. Or consider the following snippet of code:

```
1.  string input = "";
2.  cout<<"Hello"<<endl;
3.  while (input != "Goodbye") {
4.      getline(cin, input);
5.      if (input == "Are you conscious?")
6.         cout<<"Yes"<<endl;
7.      else if (input == "Are you a cute leetle kitten?")
8.         cout<<"Yes, my eyes are blue and I cry 'Mew mew mew'."<<endl;
9.      else if (input == "Goodbye")
10.        cout<<"Goodbye"<<endl;
11.     else
12.        cout<<"Nice weather for the time of year."<<endl;
13. }
```

A computer running this code will claim that it is conscious. It will also claim that it is a cute leetle kitten. Neither claim is convincing.

My wife is a *zombie*. She hides from light and shuffles home from work with dead eyes, drinks in the pub with dead eyes, makes love with dead eyes. Her physical body is *not* associated with a bubble of experience. Her zombie statements about 'consciousness' are not descriptions of a bubble of experience. They are just empty sounds produced by biochemical processes.

My wife is a professional phenomenologist. She says many things that appear to be descriptions of a bubble of experience. I cannot directly observe her lack of consciousness, so how can I prove that she is a zombie? How can I prove that other people's bodies are *really* associated with bubbles of experience? This is the traditional problem of other minds.

To scientifically study consciousness we need a physical system that is associated with consciousness. Since it is impossible to prove that particular physical systems are conscious, we have to set aside philosophical worries about solipsism and zombies and *assume* that one or more physical systems are actually conscious. I will do this by introducing the concept of a platinum standard system:

D4. A *platinum standard system* is a physical system that is assumed to be associated with consciousness some or all of the time.[12]

The term 'platinum standard system' is a reference to the platinum-iridium bar that was the first working definition of a metre.[13] Other objects were directly or indirectly compared to this platinum-iridium bar to measure their length. The length of this bar could not be checked because it was *defined* to be one metre long: when this bar expanded, everything else contracted.[14]

Platinum standard systems are the starting point for consciousness science. Consciousness is simply assumed to be associated with these systems. When we have identified the relationship between consciousness and the physical world in platinum standard systems, we can use this knowledge to make inferences about the consciousness of other systems.

I was awarded a grant to study consciousness and ordered a platinum standard system from the supplier. It was delivered yesterday. I poured myself a coffee and strolled over to inspect it. A decent enough specimen with a bushy red beard, around 2 m tall. It made angry noises and rattled the bars of its cage. I prodded it with a stick and topped up its bowl of brown nuggets.

The supplier states that this system is associated with a bubble of experience. They are a reputable firm, so I have confidence in their claim. On the first day of our experiments we strapped the platinum standard system into a chair, held up a red apple and asked it for a c-report. It stated that it was conscious of a red apple. A promising start, but it had a crafty look in its eye—it might have been *lying*. Or its consciousness might be a delusional world that is completely *disconnected* from its behaviour. I was assured that this system shipped with a bubble of experience, but the supplier did not guarantee that I would be able to use c-reports to measure its bubble of experience.

Scientists studying consciousness need to measure consciousness. While a platinum standard system's c-reports can be cross-checked for consistency, there is no ultimate way of establishing whether they are correct. Since c-reports are the *only* way in which consciousness can be measured, it has to be explicitly assumed that c-reports from a platinum standard system co-vary with its consciousness:

A1. During an experiment on consciousness, the consciousness associated with a platinum standard system is functionally connected to the platinum standard system's c-reports.

A functional connection between consciousness and c-reports is a deviation from statistical independence—not necessarily a causal connection.[15]

A1 captures the idea that our consciousness is connected to our c-reports. When our consciousness changes, our c-reports change. This assumption does not specify the amount of functional connectivity between consciousness and c-reports, which will vary with the type of c-reporting. A1 is also explicitly limited to experiments on consciousness.[16]

Outside of experiments on consciousness it is possible that a system's consciousness could be disconnected from its behaviour. Information gathered by consciousness experiments could be used to make inferences about the presence of consciousness in these situations. It could also be used to make deductions about the consciousness of systems that are not platinum standards, such as brain-damaged patients (see Section 9.2).

I contact the supplier. They issue me with a certificate that guarantees that their platinum standard systems' c-reports are functionally connected to their consciousness (A1). We resume our experiments and identify a neural firing pattern that always occurs when the platinum standard system is conscious, and never occurs when it is not conscious. We have found the correlates of consciousness! We write up the results and submit our paper for publication.

The paper is rejected. We are devastated and enraged. One reviewer argues that our platinum standard system could have several consciousnesses. The second reviewer suggests that its bubble of experience could have features that are impossible to c-report under any circumstances. The third reviewer points out that it might be conscious when it is not c-reporting—it would just be unable to remember or report its consciousness. At best we have identified a correlate of part of its consciousness, not a true correlate of consciousness.

The systematic study of consciousness will be difficult or impossible if platinum standard systems are potentially associated with ghostly ecosystems of unreportable consciousnesses, or if many aspects of consciousness cannot be c-reported. Scientific studies have to assume that this is not the case:

> **A2.** During an experiment on consciousness *all* conscious states associated with a platinum standard system are available for c-report and all aspects of these states can potentially be c-reported.[17]

This assumption states that every aspect of all of the conscious states that are associated with a platinum standard system can potentially be c-reported, even if they are not actually reported during an experiment.[18] So we can use a variety of c-reports to extract a complete picture of a platinum standard system's consciousness (see Section 4.8).[19]

A2 is incompatible with panpsychism.[20] If all matter is conscious all the time, then c-reports cannot be used to measure all of a platinum standard system's consciousness. If panpsychism was true, an apparently unconscious brain that was c-reporting zero consciousness would be associated with a bubble of experience.[21]

4.4 Pinning Consciousness to the Physical World

When I was a lad my father shone 700 nm light into my eyes and said 'Red ... red ... *red.*' My mother shone 450 nm light into my eyes and said 'Blue ... blue ... *blue.*' At a later point in time I c-report that there is a red patch in my bubble of experience. To make this report I use the association that I have learnt between an experience and a word. When I say that I am conscious of the red patch I am saying that I am having approximately the *same* colour experience that I had when I learnt the word 'red'. The incoming electromagnetic waves have activated the same brain areas that were activated when I learnt the word 'red' as a child, which presumably are associatively linked to particular language or conceptual areas. My description *of* my conscious experience is a comparison *with* earlier experiences.[22]

We are sitting in a bare whitewashed room. A human ear is on the table in front of us. The colour of the torn edge of the ear is similar to the colour that I experienced when my father shone 700 nm light into my eyes. I make a c-report: 'I am experiencing the colour red.' The colour of the torn edge of the ear in your bubble of experience is similar to the colour that you experienced when your father shone 700 nm light into your eyes. You make a c-report: 'I am experiencing the colour red.'

We both report that we are experiencing 'red', so we are apparently having the same conscious experience. But what if the colour produced by 700 nm electromagnetic waves in my bubble of experience is completely *different* from the colour produced by 700 nm electromagnetic waves in your bubble of experience? We have learnt the same mapping between incoming electromagnetic wave frequencies and colour names, so we will both make identical reports about the electromagnetic waves that we are exposed to, but nothing guarantees that these reports correspond to identical colour experiences. This is the classic problem of colour inversion, which is illustrated in Figure 4.1.[23]

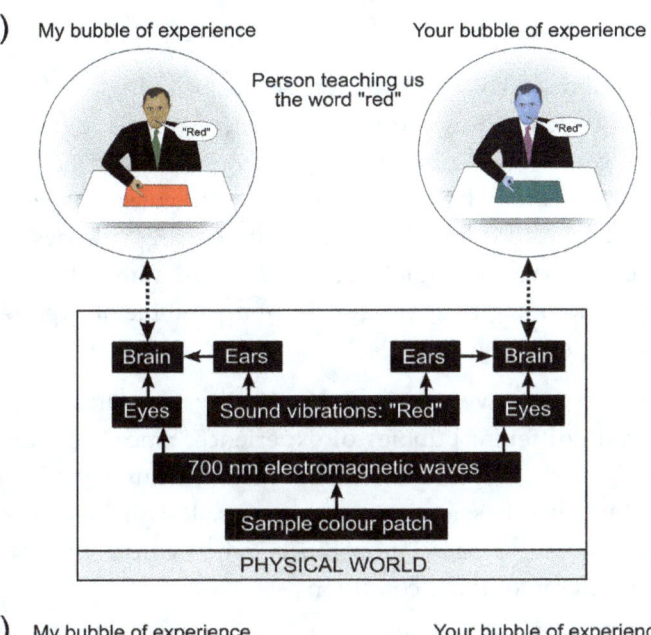

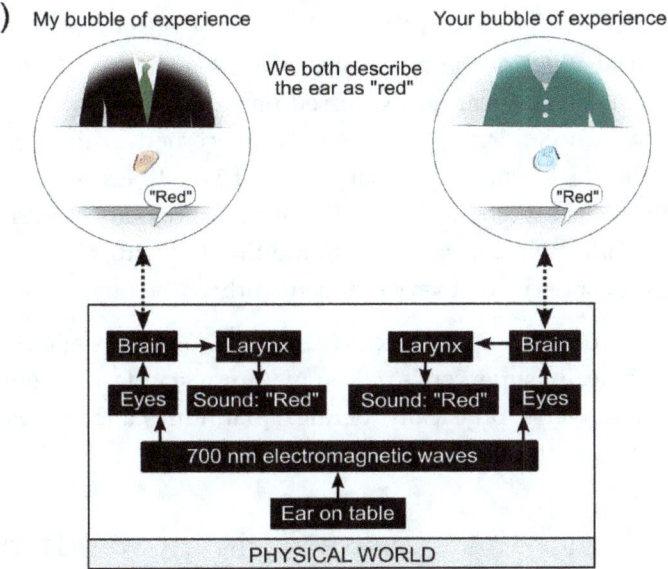

Figure 4.1. Problem of colour inversion. a) A person teaches us the word 'red' by pointing to a coloured patch and making the sound 'red'. Your colours are inverted relative to mine, so my red is your turquoise, and so on. We both learn to associate the colour that we experience with the sound 'red'. b) We observe a severed ear on a table. The colour of the torn edge of the ear is similar to the colour that we experienced when we learnt the word 'red', so we both report that we are experiencing the colour red. The colours in our bubbles of experience are very different, but there is no way of detecting this in our external behaviour. Image © David Gamez, CC BY 4.0.

In the standard colour inversion scenario a single set of colours is linked in different ways to electromagnetic waves. Our bubbles of experience could also contain completely different sets of 'colours' that have no overlap between them. Or our consciousnesses could be different in more radical ways—different geometries, different experiences of space and time, differences that I am unable to imagine because I cannot imaginatively transform my bubble of experience into these other states.

In these scenarios two systems in similar physical states are associated with radically different bubbles of experience. Since they are making the same c-reports the differences between their bubbles of experience will not show up in scientific experiments. It will be impossible to systematically study the relationship between consciousness and the physical world under these conditions.

To address this issue scientists studying consciousness have to assume that identical states of the physical world are associated with identical conscious states. This can be expressed using the philosophical concept of supervenience.[24] Since we are only concerned with a pragmatic approach to the science of consciousness, it is not necessary to assume that consciousness logically or metaphysically supervenes on the physical world. We just need to assume that the natural laws are such that consciousness cannot vary independently of the physical world:

> **A3.** The consciousness associated with a platinum standard system nomologically supervenes on the platinum standard system. In our current universe, physically identical platinum standard systems are associated with indistinguishable conscious states.

4.5 Which Systems are Platinum Standards?

It is not known when consciousness emerges in the embryo or infant.[25] We do not know whether birds or cephalopods are conscious.[26] Brain-damaged people can inaccurately report their consciousness.[27] No-one knows whether computers are capable of consciousness. We try and fail to use our imagination to decide whether consciousness is present in these systems.

I am an adult. I can smoke, drive and vote. Ten doctors claim that my brain is functioning normally. My brain does not contain unusual chemicals that might affect its operation. I am certain that *this* normally functioning adult human brain is associated with consciousness some of the time. If consciousness supervenes on the physical world (A3), then similar brains will be associated with similar consciousness:

A4. The normally functioning adult human brain is a platinum standard system.[28]

The normally functioning adult human brain is the only system that we confidently associate with consciousness. At a later point in time we might make further assumptions that extend the number of platinum standard systems. For example, we might assume that the red nodules on the genitals of an alien race are platinum standard systems.[29]

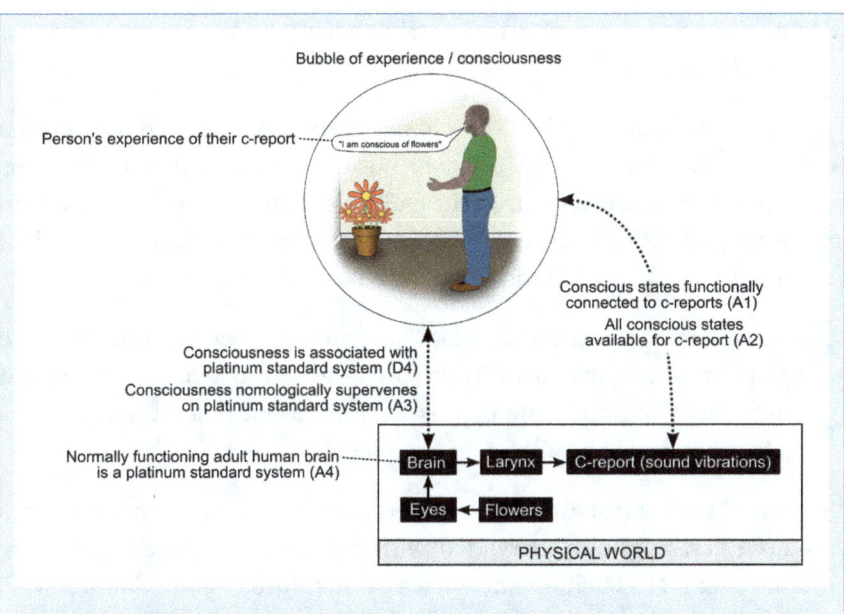

Figure 4.2. Some of the definitions and assumptions that are required for scientific experiments on consciousness. The normally functioning adult human brain is a platinum standard system (A4), which is associated with consciousness (D4). Consciousness nomologically supervenes on the platinum standard system (A3) and all of it can be c-reported (A1, A2). Image © David Gamez, CC BY 4.0.

The science of consciousness is limited by the set of systems that we assume to be platinum standards. It is a science of the relationship between consciousness and platinum standard systems. Many relationships between consciousness and the physical world might not appear in normally functioning adult human brains. This would reduce the accuracy of our deductions about consciousness in non-human systems (see Section 9.2).

Some of the definitions and assumptions that have been introduced so far are illustrated in Figure 4.2.

4.6 The Correlates of a Conscious State

Yesterday I lost consciousness in the street. My body lay crumpled on the concrete. Insects crawled over my face. Cappuccino-carrying commuters stepped over me on the way to the office. My body was just a thing—a part of the physical world that was not associated with a bubble of experience.

I am conscious now. The state of my brain now is different from the state of my brain when I lay unconscious on the street. If consciousness supervenes on the physical world (A3), something must be present in my brain now that is absent when consciousness is absent. This is a *correlate of consciousness*, which is defined as follows:

> **D5.** A *correlate of conscious state* is a minimal set[30] of one or more spatiotemporal structures in the physical world. This set is present when the conscious state is present and absent when the conscious state is absent. This will be referred to as a *CC set*.[31]

'Spatiotemporal structures' is a deliberately vague term that captures anything that might be correlated with consciousness, such as activity in brain areas, electromagnetic waves or quantum events. Chapters 6–8 discuss some of the spatiotemporal structures that might be members of CC sets.

Correlates defined according to D5 will be associated with consciousness wherever they are found.[32] Suppose CC sets only contain electromagnetic wave patterns. When a particular electromagnetic wave pattern occurs in your brain, you are immersed in a particular bubble of experience. When the electromagnetic wave pattern is absent, you have a different bubble of experience or no consciousness at all. None of the other types of spatiotemporal structure in your brain have any effect on your bubble of experience.

I distract you with a soft toy: 'Here reader, look at this, look... look... look at Teddy.' While you are playing with its ears I extract your brain from your skull and keep it alive in a jar. I provide stimulation patterns that mimic the sensory-motor responses of your discarded body. I ensure that the electromagnetic wave pattern in your brain is identical to the one that was present when you were playing with Teddy. This is associated with a bubble of experience in which you are playing with Teddy, so you continue to have this experience.

I discard your brain's biological tissue and replace it with silicon chips that are programmed to produce the same pattern of electromagnetic waves. You remain contentedly unaware of what is going on and continue to play with Teddy's ears in your bubble of experience. Suppose that the same pattern of electromagnetic waves occurs by chance when I drop my phone. This will also be associated with a bubble of experience in which you are playing with Teddy's ears.

Definition D5 enables me to state assumption A3 more precisely:

A3a. The bubble of experience that is associated with a CC set nomologically supervenes on the CC set. In our current universe, physically identical CC sets are associated with indistinguishable conscious states.

A correlation between A and B is the same as a functional connection between A and B—they are different ways of stating that A and B deviate from statistical independence.[33] So a CC set can be described as a set of spatiotemporal structures that is *functionally connected* to a conscious state. This way of describing the relationship between consciousness and the physical world will play a role in what follows, so it will be formally stated as a lemma:

L1. There is a functional connection between a conscious state and its corresponding CC set.[34]

4.7 A Causal Relationship between Consciousness and the Physical World?

> A science that invokes mental phenomena in its explanations is presumptively committed to their causal efficacy; for any phenomenon to have an explanatory role, its presence or absence in a given situation must make a difference—a *causal difference*.
>
> Jaegwon Kim, *Mind in a Physical World*[35]

I was looking for love on the Internet. ButiDD's profile looked promising: witty lines, sexy curves, hot pics. We arranged a date on Friday 5 August 2005 at 15:00 in a cafe on Hampstead Heath. When we met there was no chemistry. Conversation ground to a halt. I ate my cake. To cut through the boredom and silence I remarked 'I am conscious of a sweet taste in my mouth.' These sound vibrations led, through a complex chain of causes and effects, to Hurricane Katrina.

C-reports have physical effects. Speech vibrates the air, writing makes marks, gestures depress buttons and pull levers. These physical effects lead to further chains of causes and effects, which can be amplified into a hurricane or dissolve into background noise. Consciousness appears to be the source of c-reports, so it is natural to assume that it is the sort of 'thing' that can cause effects in the physical world.

A clearer definition of causation will help us to understand the relationship between consciousness and c-reports. First I will distinguish between conceptual and empirical theories of causation.[36] *Conceptual* theories of causation elucidate how we understand and use causal concepts in our everyday speech. *Empirical* theories of causation explain how causation operates in the physical world—by reducing it to the exchange of physically conserved quantities, such as energy and momentum, or linking it to physical forces.[37]

Conceptual analyses of causation are popular in philosophy, but it is difficult to see how our use of 'causation' in everyday speech can help us to understand the causal interactions in the brain's neural networks and the relationship between consciousness and the physical world.

Empirical theories of causation can precisely identify causal events and exclude cases of apparent causation between correlated events. They can easily relate the causal laws governing macro-scale objects, such as cars and people, to the micro-scale interactions between molecules, atoms and quarks. Empirical theories of causation are a much more appropriate starting point for studying the causal relationships between consciousness and c-reports.

A detailed discussion of the advantages and disadvantages of different theories of empirical causation is beyond the scope of this book, but it will be easier to analyze the c-reporting of consciousness with a concrete theory in mind. For this purpose I will use Dowe's theory of empirical causation. This is the most fully developed conserved quantities approach and it has the following key features:[38]

- A *conserved quantity* is a quantity governed by a conservation law, such as mass-energy, momentum or charge.
- A *causal process* is a world line[39] of an object that possesses a conserved quantity.
- A *causal interaction* is an intersection of world lines that involves the exchange of a conserved quantity.

This account of causation will be referred to as *e-causation*. The framework developed in this book relies on there being *some* workable theory of e-causation, but it does not depend on the details of any particular account. If Dowe's theory is found to be problematic, an improved version can be substituted in its place.[40]

A car moves along a road at 5 m/s and knocks a fat man down (Figure 4.3a). In this e-causal interaction energy-momentum is transferred from the car to the man. This macro-scale e-causal interaction can be reduced down to the micro-scale e-causal interactions between the physical constituents of the car and man, in which atoms in the car's bumper pass energy-momentum to atoms in the man's legs (Figure 4.3b).

We can distinguish between true and false causes of this event. The car's engine temperature is a macro-scale property of the physical world that moves along at the same speed as the car and also collides with the man (Figure 4.3c). However, the macro property of engine temperature does not exchange energy-momentum with the man, so it does not e-cause him to fall down, although it can e-cause other macro-scale effects, such as skin burns. Similar e-causal accounts can be given of the laws of other macro-scale sciences, such as geology, chemistry and biology.[41]

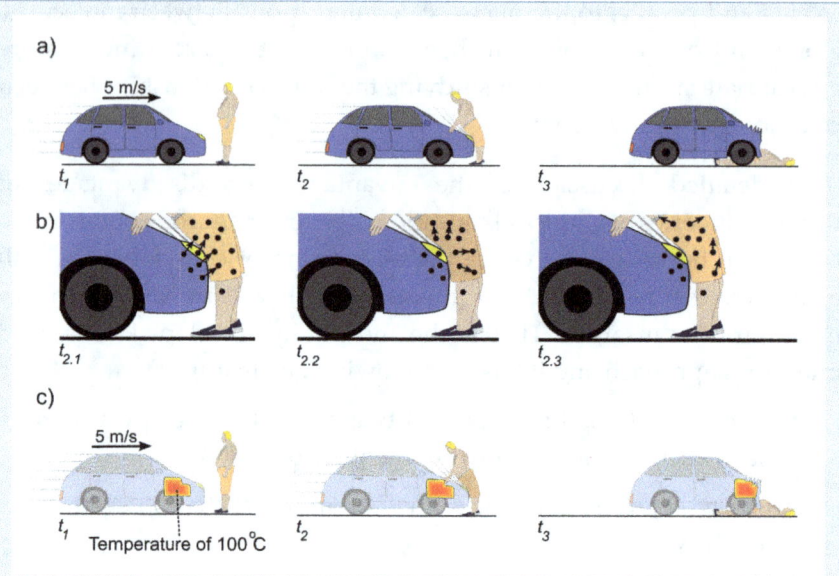

Figure 4.3. The relationship between macro- and micro-scale e-causal events. a) A car moving at 5 m/s collides with a fat man and knocks him down. This is a macro-scale e-causal event in which the car passes energy-momentum to the man. b) The macro-scale e-causal interaction between the car and man can be reduced down to the micro-scale exchanges of energy-momentum between atoms in the car and man. c) The temperature of the car's engine is a macro-scale property that moves at 5 m/s and collides with the man. The engine temperature exchanges a small amount of energy-momentum with the man in the form of heat, but not enough to e-cause him to fall down. Image © David Gamez, CC BY 4.0.

It is generally assumed that the amount of energy-momentum in the physical universe is constant (as long as the reference frame of the observer remains unchanged). When part of the physical world gains energy-momentum, this energy-momentum must have come from elsewhere in the physical universe. It is also generally assumed that the net quantity of electric charge in the universe is conserved. If part of the physical world gains electric charge, another part of the physical world must have lost charge or there must have been an interaction in which equal quantities of positive and negative charge were created or destroyed. Similar arguments apply to other physically conserved quantities, which leads to the following assumption:

A5. The physical world is e-causally closed.

According to A5, any change in a physical system's conserved quantities can in principle be traced back to a set of physical e-causes that led the system to gain or lose those conserved quantities at that time.

In everyday language we say that a person reports or describes their consciousness. This might naively be interpreted as the idea that consciousness directly or indirectly alters the activity of the brain's speech areas, sending spikes to the larynx that lead to sound vibrations in the air.

The problem with this naive picture is that consciousness could only e-cause a chain of events leading to a c-report if it could pass a physically conserved quantity, such as energy-momentum or charge, to neurons in the c-reporting chain—for example, if it could push them over their threshold and cause them to fire.[42] If the physical world is e-causally closed (A5), then a conserved quantity could only be passed from consciousness to a brain area if consciousness is a physical phenomenon, i.e. if consciousness *is* the correlates of consciousness.[43]

Consciousness *is* the correlates of consciousness *if* physicalism is correct. But it would be premature and controversial to base the scientific study of consciousness on this assumption. It is also absurd to claim that a bubble of experience *is* a pattern of invisible wave-particles. It would be much better to find a way of measuring consciousness that does not depend on the assumption that physicalism is true.

I have assumed that a conscious state is functionally connected to a CC set (L1) and that c-reports are functionally connected to consciousness (A2). To fully account for the measurement of consciousness we need an e-causal connection between CC sets and c-reports. This can be addressed by introducing a further assumption that fits in naturally with the current framework:

A6. CC sets e-cause a platinum standard system's c-reports.

This states that the correlates of consciousness are the first stage in a chain of e-causation that leads to c-reports about consciousness.[44,45] It can be difficult to measure e-causation, so in some circumstances A6 can be substituted for the weaker assumption:

A6a. CC sets are effectively connected to a platinum standard system's c-reports.[46]

Assumption A6 is illustrated in Figure 4.4.

By themselves A6 and A6a do not say anything about the strength of the relationship between CC sets and c-reports. There could be a very weak e-causal chain leading from a CC set to a c-report, which could primarily be driven by unconscious brain areas. The weaker the connection between CC sets and c-reports, the more experiments will be required to identify CC sets.

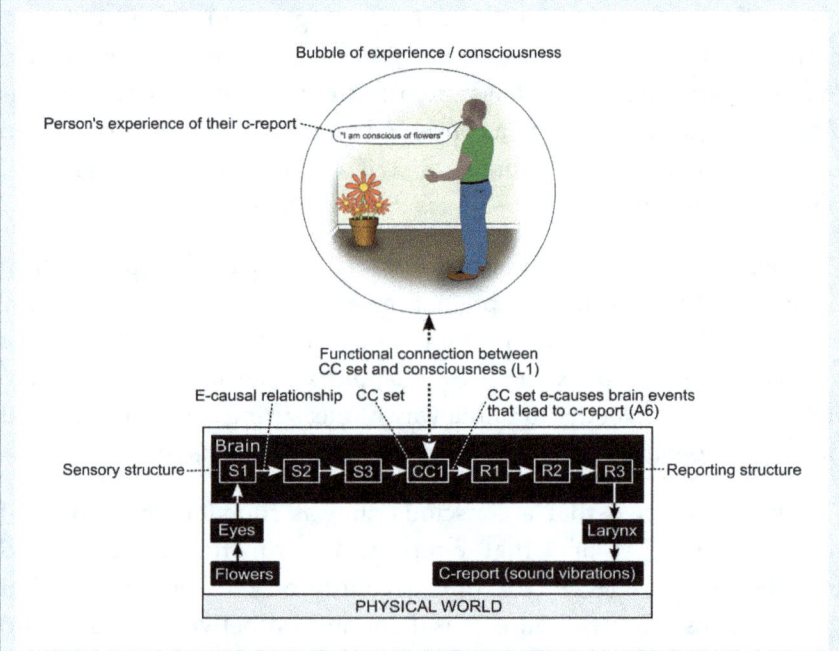

Figure 4.4. Assumptions about the relationship between CC sets, consciousness and first-person reports. The labels S1, CC1, R1, etc. refer to any kind of spatiotemporal structure in the brain, such as the activation of a brain area, neural synchronization, electromagnetic waves, quantum events, and so on. They are only illustrative and not intended to correspond to particular anatomical paths or structures. An e-causal chain of sensory spatiotemporal structures, S1-S3, leads to the appearance of a spatiotemporal structure, CC1, that is functionally connected to consciousness. In this example the contents of consciousness are determined by sensory events, but in principle they could be independent of S1-S3—for example, if the subject was dreaming. CC1 is assumed to be the first stage in an e-causal chain of spatiotemporal structures, R1-R3, that lead to a verbal description of consciousness. Image © David Gamez, CC BY 4.0.[47]

4.8 The Limits of C-Reporting

You are looking at your reflection in a mirror. You see greying hair, burst capillaries, lengthening deepening lines. A tired sad sagging face. Your youth has gone. You will die soon. A sense of helpless fatality washes over you. You imagine how your face will look in the grave, under the wet earth, your empty eye sockets staring blankly at blackness, while the world rolls along and your existence fades away without trace.

Stick up your thumbs and interlace your fingers. Extend your arms to their full length in front of you. Look directly at your thumbnails. The area covered by your thumbnails is the high resolution part of your visual field. The rest is low resolution. When you look at your nose in the mirror only sketchy information is coming in from your gold earrings and beard. You cannot detect substantial changes that occur outside of the high resolution area.[48]

The limited extent of our high resolution vision is not a problem in daily life. When we require more information about a feature of our environment we make a rapid eye movement (known as a saccade) to bring this feature into high resolution vision. As you look in the mirror you are moving your eyes every ~200 ms. You inspect the pores on your nose, flick across to your left cheek, look up at your eyebrow, and so on.[49]

Depressed you pluck out a protruding hair. You squeeze a painful spot and wipe a stain from the mirror. A chewed-up cabbage leaf is trapped between your teeth. You remember your dental appointment tomorrow.

Your consciousness changes several times per second.[50] As you look at your face in the mirror you are receiving fresh sensory information from your eyes and body and attending to different senses (moving from vision, to touch, to audition, etc.). You are shifting between past, present and possible futures: between memory, perception and imagination.

Describe your consciousness *now*. You were in a reverie—try again. The clock reads 22:59:50.874. Describe your consciousness when it changes to 23:00:00.000. Get ready ... *now*.

When you started to describe your consciousness you were alert and speaking coherently. This external behaviour was a c-report of a high level of consciousness. Immobility and incoherent mumbling would have been a c-report of a low level of consciousness.

When the clock changed to 23:00:00.000 you started to describe your consciousness *in natural language*. But your consciousness changed when you uttered the first word—it became consciousness of that word. The consciousness that you had at 23:00:00.000 vanished when you started to describe it. Natural language is too slow to c-report consciousness in real time.

Ok, try a different strategy. The clock reads 23:01:46.340. Describe your consciousness when it changes to 23:02:00.000. Get ready ... *now*.

This time you tried to *remember* your state of consciousness at 23:02:00.000. You converted an online bubble of experience into an offline bubble of experience. This memory preserves some washed-out unstable information about the visual consciousness that you had at 23:02:00.000. It holds little detail about the sounds, smells, tastes and body sensations that were in your bubble of experience at that moment. Your memory is also fragile—it is not like a computer file. As you describe your memory of your consciousness at 23:02:00.000 it becomes contaminated with details that came before or after the moment that you are trying to remember.

One more attempt. The clock reads 23:05:51.087. Describe your consciousness when it changes to 23:06:00.000. Get ready ... *now*.

You could not accurately remember what your nose looked like at 23:06:00.000. So when I asked you to describe your consciousness you moved your eyes to look at your nose. You used the *world as external memory*.[51] Your face is pretty stable, so perhaps you could use this method to generate a complete description of your consciousness at 23:06:00.000. However, this would not provide a description of your consciousness at 23:06:00.000: it would be a description of a series of moments of consciousness in which different aspects of your face enter the high resolution part of your bubble of experience. When you moved your eyes to obtain information about your nose, your consciousness at 23:06:00.000 was replaced with a new bubble of experience in which you were 'zoomed in' on your nose.

4. The Measurement of Consciousness

We cannot accurately describe a state of our consciousness. Natural language is too slow and vague. Our memory does not store enough details. Our consciousness at a given moment cannot be reconstructed by re-accessing information from our environment.[52]

Fixate your eyes on the small cross in the centre of the screen. Rest your index finger on the button in front of you. When I say 'now' I want you to press the button if there is a small red square in the bottom left hand corner of your visual field. Get ready ... *now*.

Under controlled experimental conditions I can extract a small amount of accurate information about a specific aspect of your consciousness at a given time. The details of the measurement are set by the experimental conditions. The subject is only required to answer a simple yes/no question, without any need for memory or natural language.

This measurement method has the limitation that a subject can only answer one or two yes/no questions before their consciousness changes. This problem can be partly overcome by resetting their consciousness after each measurement. We can then use a large number of high precision probes to obtain a detailed measurement of one state of a subject's consciousness.[53]

As an example, consider an experiment that measures a subject's visual consciousness. To begin with the subject is asked to fixate on a cross on a screen. When they are looking at the cross it is replaced with a picture that remains on the screen for 200ms. This is long enough to ensure that the subject becomes conscious of the picture, and short enough to prevent them from moving their eyes while they are looking at it. The brief exposure attracts their attention—reducing the chance that their visual consciousness is remembering or imagining something else. The subject's fixation on the cross ensures that their bubble of experience contains the same part of the picture each time. When the subject's visual consciousness is put into this state one aspect of it can be measured with high precision. Repetition of this procedure can be used to progressively build up a detailed description of this state of consciousness.[54]

High precision measurement combined with the resetting of consciousness under experimental conditions is the most promising method for obtaining detailed descriptions of consciousness. But there are limits to the types of consciousness that can be reset, and a subject's consciousness cannot be put into exactly the same state each time. These problems will reduce our ability to obtain detailed accurate measurements of consciousness.[55]

4.9 Formal Descriptions of Consciousness (C-Descriptions)

> At present we are completely unequipped to think about the subjective character of experience without relying on the imagination—without taking up the point of view of the experiential subject. This should be regarded as a challenge to form new concepts and devise a new method—an objective phenomenology not dependent on empathy or the imagination. Though presumably it would not capture everything, its goal would be to describe, at least in part, the subjective character of experiences in a form comprehensible to beings incapable of having those experiences.
>
> Thomas Nagel, What Is It Like to Be a Bat?[56]

Alice and Bob measure your consciousness at 14:02:00.050 and submit written reports of the results. Alice's report contains several thousand words of natural language, similar to the work of Husserl, Heidegger and Merleau-Ponty. Bob's report contains natural language descriptions of the experimental probes that he ran on your consciousness. It is written in the style of a methods section in a paper on experimental psychology. When you read either of these reports you are satisfied that they are a complete and accurate description of your consciousness at 14:02:00.050.

This verification process is inexact. It relies on an inaccurate memory of your state of consciousness. It is far from a complete validation. But it will have to do—it is all we can do.[57]

Formal descriptions play an important role in science. We have formal descriptions of many aspects of the physical world (mass, charge, voltage, magnetic field, etc.) that can be used to generate testable predictions. The Earth and Sun can be described as point masses of 5.97 × 10²⁴ kg and 1.99 × 10³⁰ kg. We can use this description of the Earth and Sun to predict the gravitational force between them (by substituting the masses for m_1 and m_2 in Newton's equation $F=Gm_1m_2/r^2$).[58]

Scientific theories of consciousness will eventually use mathematics to map between descriptions of consciousness and descriptions of the physical world (see Section 5.5). This will enable us to make strong testable predictions about the conscious state that is associated with a physical state. This will only become possible when consciousness can be described in a formal way that can be manipulated by algorithms and mathematical equations. This will be referred to as a c-description:

D6. A *c-description* is a formal description of a conscious state.

C-descriptions must be compatible with mathematics and they must be applicable to both human and non-human consciousness. We will have to develop methods for converting c-reports into c-descriptions and vice versa.

Natural language cannot be used for c-descriptions. It is vague, ambiguous, highly compressed and context dependent. Natural language descriptions of consciousness are difficult to analyze with algorithms and it is not obvious how they can be integrated with mathematical equations. Natural language also cannot be used to describe the consciousness of non-human systems, such as infants, bats or robots.[59]

C-descriptions could be written in a markup language, such as XML or LMNL. Markup languages are more precise and tightly structured than natural language, and they can be read by both humans and computers. They can capture complex nested hierarchies, which would enable them to describe the relationships between different parts and aspects of a conscious state.[60]

Mathematics could be used for c-descriptions. For example, Balduzzi and Tononi have suggested how conscious states can be described using high dimensional mathematical structures.[61] Other mathematical techniques could be used to describe consciousness, such as category theory or graph theory.

An adequate c-description format is essential for the scientific study of consciousness. C-descriptions are at a very early stage of development and we are only just starting to explore solutions.

4.10 Summary

Scientists studying consciousness need to accurately measure conscious states. Consciousness is measured through first-person reports (c-reports), such as speaking or body gestures, which cannot be independently checked. This raises the philosophical problems of zombies, solipsism, colour inversion and the causal relationship between consciousness and the physical world. These problems cannot be solved. They can be neutralized by making assumptions that guarantee that consciousness can be accurately measured. The results of the science of consciousness can then be considered to be true given these assumptions.

I started by assuming that consciousness is functionally connected to first-person reports (A1). I then assumed that everything about a conscious state can be reported during an experiment and that there are no ghostly consciousnesses floating around that cannot be reported (A2). I handled colour inversion scenarios by assuming that consciousness supervenes on the brain (A3, A3a). First-person reporting does not break the causal closure of the physical world (A5) because reports about consciousness are e-caused by the correlates of consciousness (A6). All of these assumptions apply to systems that are assumed to be conscious (platinum standard systems) during experiments on consciousness. I assumed that normally functioning adult human brains are platinum standard systems (A4).

Consciousness cannot be described in real time using natural language, so we have to use experimental probes to measure specific aspects of a conscious state, and then reset the state and apply more probes until a complete measurement is obtained. The final output of a measurement of consciousness should be a c-description written in a tightly structured formal language, such as category theory or XML, that will support the development and testing of mathematical theories of consciousness.

This chapter also introduced the concept of a CC set. A CC set is a set of spatiotemporal structures in the physical world that is correlated with a conscious state (D5). The science of consciousness attempts to develop mathematical theories that describe the relationship between CC sets and conscious states. This is covered in the next five chapters.

5. From Correlates to Theories of Consciousness

5.1 Measurement of the Physical World

Randy is an elephant who lives at the bottom of my garden. Six blind men often come round to feel Randy. Sometimes I like to measure Randy. To measure his height, I compare my conscious experience of Randy with my conscious experience of a stick that has been calibrated against the distance light travels in a vacuum during 1/299,792,458 seconds.[1] The ratio between Randy and the stick is his height in metres. Randy is three sticks (three metres) high (see Figure 5.1).[2]

The science of consciousness studies the relationship between consciousness and the physical world. It measures consciousness, measures spatiotemporal structures in the physical world and attempts to identify minimal sets of spatiotemporal structures (CC sets) that are linked to conscious states.

Physical objects do not directly appear in our bubbles of experience. We do not directly perceive their mass, chemical composition or size. To measure a property of a physical object I cause it to interact with another physical object that has been calibrated in some way and observe this interaction in my bubble of experience. This typically results in a number. Eddington describes this process:

> Let us then examine the kind of knowledge which is handled by exact science. If we search the examination papers in physics and natural philosophy for the more intelligible questions we may come across one beginning something like this: "An elephant slides down a grassy hillside…" The experienced candidate knows that he need not pay much attention to this; it is only put in to give an impression of realism. He reads on: "The mass of the elephant is two tons." Now we are getting down to business; the elephant fades out of the problem and a mass of

two tons takes its place. What exactly is this two tons, the real subject matter of the problem? It refers to some property or condition which we vaguely describe as "ponderosity" occurring in a particular region of the external world. But we shall not get much further that way; the nature of the external world is inscrutable, and we shall only plunge into a quagmire of indescribable. Never mind what two tons refers to; what is it? How has it actually entered in so definite a way into our experience? Two tons is the reading of the pointer when the elephant was placed on a weighing-machine. Let us pass on. "The slope of the hill is 60°." Now the hillside fades out of the problem and an angle of 60° takes its place. What is 60°? There is no need to struggle with mystical conceptions of direction; 60° is the reading of the plumb-line against the divisions of a protractor. Similarly for the other data of the problem.[3]

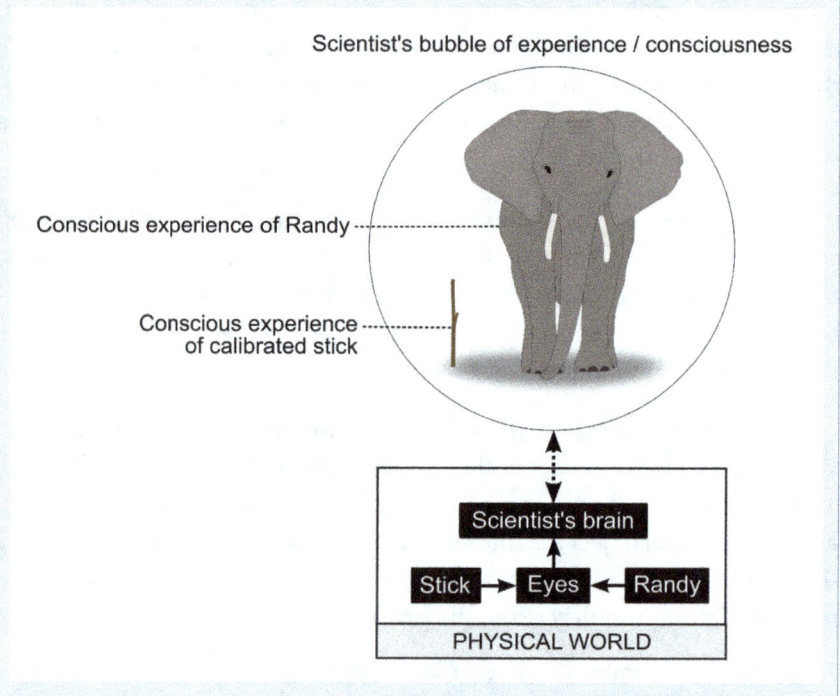

Figure 5.1. The measurement of an elephant's height in a scientist's bubble of experience. The scientist compares Randy with a stick that has been calibrated against the distance light travels in a vacuum during 1/299,792,458 seconds. The ratio between Randy and the calibrated stick is his height in metres. Randy is three sticks (three metres) high. Image © David Gamez, CC BY 4.0.

When I monitor Randy's brain activity using electrodes, my equipment compares the effect of his brain on each electrode with the effect of a standard voltage: the ratio between these effects is the electrode's voltage. A computer converts the electrode voltages into an attractive image of brain activity. Randy's colourless physical brain does not directly appear to me when I am measuring it. Within my bubble of experience I am conscious of black wires emerging from a pinkish-grey brain and a 3D display of brain activity on a computer screen.

The physical world can be measured automatically without the instruments or measured objects appearing in a bubble of experience. A robot could measure Randy with a stick and write down the result on a piece of paper.

Measurements can be processed into numbers that correspond to different properties of an object. Electrode voltages can be processed into neuron firing events, which can be processed into firing frequencies, synchronization patterns, and so on.

Measurement assigns numbers to aspects of objects, properties or events.[4] Objects, properties and events are typically described in *natural language*. When I measured Randy, 3 metres was the *height of an elephant*; 30 mV was the *membrane potential of a neuron*.

Objects, properties and events are tightly defined in physics and chemistry. For example, we have clear definitions of quarks and carbon. Physicists and chemists can state exactly what it means for a physical object to contain quarks or carbon; their instruments can reliably detect whether quarks or carbon are present in a physical object.

Context plays an important role in the description of biological objects, properties and events. Suppose I want to measure the membrane potential of a neuron. I do not use an abstract definition of a neuron to identify physical objects that are neurons—I look for a particular type of cell in the brain of an animal. The definition of a neuron only has to be precise enough to distinguish neurons from other cells in the brain. The *context* of a neuron (in the brain of an animal) is part of its definition.

A neuron is well defined inside a brain—but what exactly *is* a neuron? Does a neuron continue to be a neuron if I remove its nucleus, give it a chrome cytoskeleton and change its resting potential to 100 V? Synthetic biologists could construct a series of intermediate cases between neurons and liver cells—it would be difficult to classify the intermediate cases. Neurons are defined in a specific biological context; no formal definition exists that can unambiguously decide whether an arbitrary physical object is a neuron.

This vague definition of biological structures is a problem for consciousness science. We want to use what we know about consciousness in the brain to make inferences about the consciousness of non-biological systems. Suppose we identify neural correlates of consciousness and want to make inferences about the consciousness of synthetic neurons. This cannot be done without an unambiguous context-free definition of a neuron.

To address this problem we need formal ways of describing the spatiotemporal structures that form CC sets. These will be referred to as p-descriptions:

D7. A *p-description* is a formal description of a spatiotemporal structure in the physical world. A p-description unambiguously determines whether a spatiotemporal structure is present in a sequence of physical states.

When a spatiotemporal structure can be completely described by physics or chemistry (for example, an electromagnetic field or a molecule), the p-description is identical to the standard scientific description. We will have to find more formal context-free ways of describing biological structures that can resolve ambiguous cases. For example, we need a p-description that can determine whether an arbitrary part of the physical world contains neurons. This should not rely on the fact that neurons are found in biological creatures, and it should provide definite classifications of synthetic neurons, which lack some of the attributes of biological neurons. If the members of a CC set cannot be captured in a p-description, then we will only be able to make inferences about the consciousness of systems that are similar to platinum standards.[5]

5.2 Constraints on CC Sets

There are constraints on the spatiotemporal structures that can form CC sets. These derive from the assumptions that were introduced to measure consciousness (A1-A6) and from the requirements of scientific methodology:

C1. *The spatiotemporal structures in a CC set are independent of the observer.* My consciousness is a real phenomenon that does not depend on someone else's subjective interpretation. CC sets must be formed from objective spatiotemporal structures, such as electromagnetic waves and neuron firing patterns.

C2. *The members of CC sets are intrinsic properties.*[6] A conscious state supervenes on a CC set (A3a), so each duplicate of a CC set must be associated with an identical conscious state, regardless of the spatial and temporal context in which the duplicate appears.

C3. *A non-conscious system does not contain a CC set that is 100% correlated with a conscious state.*[7] If A and B are 100% correlated, then A cannot occur without B. If a CC set is 100% correlated with a conscious state, then all brains that contain that CC set will be conscious.

C4. *CC sets e-cause c-reports during consciousness experiments (A6).*[8] It is not necessary for every member of a CC set to e-cause c-reports. But some parts or aspects of the CC set must e-cause them. So when I say 'I am conscious of a green tomato', this c-report can be traced back to the CC set that e-caused it, which is functionally connected to a bubble of experience in which there is a green tomato.

A set of spatiotemporal structures that does not conform to these constraints cannot be a correlate of a conscious state.

5.3 Pilot Studies on the Correlates of Consciousness

We are in a beastly state of ignorance about the relationship between consciousness and the physical world. We have no idea which spatiotemporal structures form CC sets. Our intuitions are useless. We have to start with the assumption that everything in a platinum standard system that conforms to the constraints is a potential member of a CC set.

How can we reduce our ignorance? We can carry out pilot studies. We can attempt to identify the CC set that is associated with a particular conscious state.[9]

Briony is an adult human with a normally functioning brain. I strap her into a chair and connect electrodes to her temples. At intervals I display a red square at the centre of her visual field, play a loud 500Hz tone and deliver an electric shock. Briony's attention is completely consumed by these stimuli. They are so compelling that the same conscious state can be induced on multiple occasions. This is conscious state c_3.

Each time c_3 is induced I ask Briony to c-report one aspect of it. On subsequent occasions she c-reports the size of the square, the colour of the tone and the shocking sensations in her body. Over time I build up a c-description of c_3 that has a tolerable amount of detail.

I want to identify the minimal set of spatiotemporal structures that is correlated with c_3. When c_3 is induced I measure the neuron activity in Briony's brain as well as the electromagnetic waves, blood movements, glia activity, and so on.[10] Some of these spatiotemporal structures could form the CC set by themselves—a pattern of neuron activity might be the sole correlate of c_3. Or a combination of spatiotemporal structures might form the CC set that is correlated with c_3. For example, a pattern of neuron activity might only be associated with consciousness when it is immersed in blood—the same neuron activity without blood would not be linked to consciousness.

I must systematically consider all possible combinations of spatiotemporal structures that could form the CC set. This will enable me to identify the spatiotemporal structures that only occur when c_3 is present. Suppose I want to demonstrate that a pattern of neuron activity, p_2, is the sole member of the CC set. I will need to measure c_3 when p_2 is present and blood is present, measure c_3 when p_2 is present and blood is absent, measure c_3 when just blood is present, and measure c_3 when neither p_2 nor blood are present. Further experiments will be required to distinguish p_2 from glia activity, cerebrospinal fluid, and so on. This methodology is illustrated in Table 5.1.[11]

Spatiotemporal Structures				Conscious States	
A	B	C	D	c_1	c_2
0	0	0	0	0	0
0	0	0	1	0	0
0	0	1	0	0	1
0	0	1	1	0	1
0	1	0	0	0	0
0	1	0	1	0	0
0	1	1	0	0	1
0	1	1	1	0	1
1	0	0	0	0	0
1	0	0	1	0	0
1	0	1	0	0	1
1	0	1	1	0	1
1	1	0	0	1	0
1	1	0	1	1	0
1	1	1	0	1	1
1	1	1	1	1	1

Table 5.1. Simple example of correlations that could exist between spatiotemporal structures in a physical system and two conscious states. It is assumed that conscious states c_1 and c_2 can occur simultaneously. The physical structures A, B, C and D could be dopamine, haemoglobin, neural synchronization, electromagnetic waves, etc. These are assumed to be the only possible features of the system. '1' indicates that a feature is present; '0' indicates that it is absent. In this example D is not a correlate of consciousness because it does not systematically co-vary with either of the conscious states. {A,B} is a set of spatiotemporal structures that correlates with conscious state c_1. {C} is a set of spatiotemporal structures that correlates with conscious state c_2.

Many pilot studies have been carried out on the correlates of consciousness. They have identified areas of the brain and features of neuron activity (for example, recurrent connections) that are potential members of CC sets.[12] Most of these pilot studies have focused on neural patterns that might form CC sets. No attempt has been made to show that neuron activity patterns form CC sets by themselves, or to demonstrate that glia, electromagnetic waves and haemoglobin are not members of CC sets.

5.4 Natural and Unnatural Experiments

Consciousness experiments are carried out on platinum standard systems. Assumptions A1-A6 enable us to measure the consciousness of platinum standard systems during these experiments.

Normally functioning adult human brains are our only platinum standard systems (A4). They change as they interact with the world and learn from their experiences. Most of these changes are part of their normal behaviour—they do not affect their status as platinum standards.

Brian's skull contains a normally functioning adult human brain (a platinum standard system). I ask him to raise his right arm. He raises his right arm. I shave off his hair and slowly smear chocolate sauce on his face. These modifications do not affect the normal functioning of his adult human brain.

I inject Brian with 5 mg of LSD. After a brief spell of bliss he goes wild, bangs his head against the wall, yells in an uncontrollable manner and claws at his face. His brain is not functioning normally. I shoot him in the head. He lies still on the laboratory floor. Blood pours out of his head. His brain is no longer a platinum standard system.

A platinum standard system must remain a platinum standard system throughout an experiment on consciousness. If it ceases to be a platinum standard system, then assumptions A1-A6 no longer hold and it becomes an open question whether we can interpret its external behaviour as a c-report of its consciousness.

Some experiments preserve a system's status as a platinum standard; other experiments transform a system into something that is not a platinum standard. This distinction will be expressed as follows:

D8. In a *natural experiment* the test system preserves its status as a platinum standard. Assumptions A1-A6 remain valid and consciousness can be measured throughout the experiment.

D9. In an *unnatural experiment* the test system is transformed into something that is not a platinum standard. A1-A6 cease to apply and we lose our ability to measure the system's consciousness.

Natural experiments preserve a system's physical integrity and normal behaviour. The system can be monitored using passive techniques, such as fMRI, EEG and electrodes.[13] These manipulations do not affect our belief that it can c-report its consciousness.

Unnatural experiments alter the physical constitution of a platinum standard system. They remove material, add unusual chemicals or replace brain parts with functionally equivalent chips. Unnatural experiments undermine our ability to measure a system's consciousness. They cannot be used to identify CC sets or to test theories of consciousness.

Suppose we replace part of a subject's brain with a functionally equivalent chip. This would not affect their behaviour—they would continue to make the same reports as before. This experiment has been put forward as a way of testing the hypothesis that functions or computations in the brain are linked to consciousness, rather than patterns in biological materials.[14]

Prior to the experiment the subject's brain was a platinum standard system and we interpreted its speech as a c-report of its conscious states. The implantation of the chip transforms the subject's brain into a freak neuro-silicon hybrid that is not a platinum standard system. We have no idea whether brains with implanted chips are associated with consciousness. Assumptions A1-A6 do not apply—we have not assumed that the external behaviour of this type of system is a c-report that can be used to measure consciousness. Similar problems occur with other unnatural experiments, such as the replacement of haemoglobin with an artificial blood substitute, the removal of glia, and so on.[15,16]

We could add brains with implanted chips to our list of platinum standard systems. This would transform an unnatural chip implantation experiment into a natural experiment. Both the original system and the transformed system would be platinum standards, so we could measure consciousness throughout the experiment.

New assumptions about platinum standard systems should not be made lightly. Pilot studies look for CC sets in the systems that are assumed to be platinum standards. A science of consciousness that studied brains with implanted chips would be very different from our current science of consciousness.[17]

It will be difficult or impossible to identify all the members of a CC set using natural experiments. The members of a CC set can only be identified by systematically varying the physical world to test the link between each combination of candidate structures and consciousness (see Table 5.1). When a combination does not occur naturally it is impossible to test its link with consciousness. So natural experiments cannot test the connection between consciousness and biological neurons, because we cannot remove biological neurons from the brain without compromising its status as a platinum standard system.

5.5 Theories of Consciousness (C-Theories)

It is possible to interpret the ways of science more prosaically. One might say that progress can '... come about in only two ways: by gathering new perceptual experiences, and by better organizing those which are available already'. But this description of scientific progress, although not actually wrong, seems to miss the point. It is too reminiscent of Bacon's induction: too suggestive of his industrious gathering of the 'countless grapes, ripe and in season', from which he expects the wine of science to flow: of his myth of a scientific method that starts from observation and experiment and then proceeds to theories [...] The advance of science is not due to the fact that more and more perceptual experiences accumulate in the course of time. [...] Bold ideas, unjustified anticipations, and speculative thought, are our only means for interpreting nature: our only organon, our only instrument, for grasping her.

Karl Popper, *The Logic of Scientific Discovery*[18]

What's the matter with consciousness, then, and how should we proceed? Early on, I came to the conclusion that a genuine understanding of consciousness is possible only if empirical studies are complemented by a theoretical analysis. [...] This state of affairs is not unlike the one faced by biologists when, knowing a great deal about similarities and differences between species, fossil remains, and breeding practices, they still lacked a theory of how evolution might occur. What was needed, then as now, were not just more facts, but a theoretical framework that could make sense of them.

Giulio Tononi, Consciousness as
Integrated Information: A Provisional Manifesto[19]

When people studied the heavens they were not seeking an infinitely long list of the planets' positions. They wanted a compact theory that could calculate the positions of the planets at an arbitrary point in time. Ptolemy developed a model based on deferents and epicycles. This was superseded by Newton's and Einstein's equations.

Pilot studies might identify the correlates of some conscious states. This would be a major scientific achievement. It would help us to find the correlates of other conscious states. It would tell us something about the consciousness of non-platinum standard systems, such as coma patients, bats and robots.

The wine of a science of consciousness will not flow from industrious gathering of data about the correlates of individual conscious states. There are an effectively infinite number of conscious states — we cannot identify the CC sets associated with each one. Instead we need a compact mathematical theory that can map physical states onto conscious states and vice versa. This will be referred to as a c-theory:[20]

D10. A *c-theory* is a compact expression of the relationship between consciousness and the physical world. A c-theory can generate a c-description from a p-description, and generate a p-description from a c-description.[21]

The role of c-theories is illustrated in Figure 5.2.

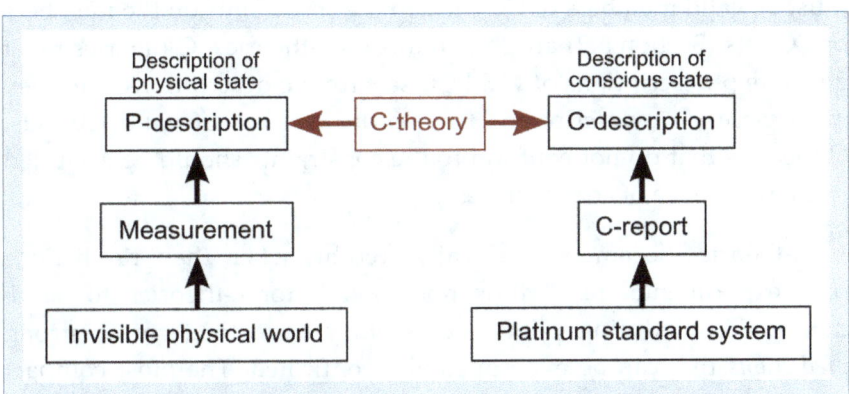

Figure 5.2. Theory of consciousness (c-theory). On the left, a measurement of the invisible physical world is converted into a formal p-description of a physical state. On the right, consciousness is measured with a c-report, which is converted into a formal c-description of a conscious state. The c-theory maps between the p-description and the c-description. Image © David Gamez, CC BY 4.0.

C-theories specify which *types* of spatiotemporal structures form CC sets. For example, neuron activity patterns, information patterns or computations might be members of CC sets. The most popular types of c-theory are covered in the next three chapters.

C-theories should be based on mathematics.[22] This is the most compact way of linking p-descriptions to c-descriptions. Philosophical theories of consciousness might inspire c-theories. But the relationship between c-descriptions and p-descriptions cannot be expressed in natural language. Natural language is too weak and vague—it cannot make strong testable predictions.

Suppose we discover a neuron whose firing rate is correlated with a bubble of experience in which there is a single point of red light. We develop a c-theory that uses the equation $ln(r)=2i$ to connect the neuron's firing rate, r, with the intensity of the conscious red light, i. This theory predicts that when the neuron fires at 7 Hz it will be associated with conscious red light that has intensity 0.97. It also predicts that the neuron will fire at 20 Hz when conscious red light occurs with intensity 1.5.[23]

C-theories map between conscious states and sets of spatiotemporal structures in the physical world. These spatiotemporal structures must be valid members of CC sets. So the constraints on the members of CC sets (Section 5.2) are constraints on c-theories. C-theories must generate p-descriptions of valid CC sets from c-descriptions, and they must generate c-descriptions from p-descriptions of valid CC sets. C-theories that do not conform to the constraints should be excluded from the science of consciousness.

C-theories become scientifically credible when their predictions pass experimental tests. It is not enough for c-theories to *match* data gathered during pilot studies—they have to generate strong *predictions* that can be experimentally confirmed. The most compact and accurate c-theory will be considered to be a correct description of the relationship between consciousness and the physical world.[24]

Our *final* c-theories will describe brute regularities in the relationship between consciousness and the physical world (see Section 3.5). As the science of consciousness progresses we are likely to develop c-theories that describe regularities which can be further decomposed into more basic relationships. It might be impossible to tell whether a c-theory describes a genuine brute regularity.[25]

Some people base c-theories on their conscious experiences.[26] But the source of inspiration of a c-theory is irrelevant to its success. We cannot directly imagine the relationship between consciousness and the invisible physical world (see Section 3.3), so the intuitive plausibility of a c-theory has no bearing on whether it is correct. C-theories stand or fall on their ability to make falsifiable predictions that pass experimental tests.

C-theories are not likely to provide intuitively satisfying explanations of the relationship between consciousness and the physical world. Since we cannot imagine the physical world, a mathematical c-theory cannot help us to make an imaginative transition from the invisible physical world to consciousness. At most a c-theory could help us to make an imaginative transition from a conscious experience of brain activity to another conscious experience (see Section 3.4).

People might use particles, forces or novel aspects of the physical world to *explain* why a particular relationship between c-descriptions and p-descriptions holds (see Section 6.3). Such explanations might be scientifically fruitful—they might help us to develop new mathematical c-theories. But they are unlikely to make a c-theory more intuitively plausible. Newton could not imagine how gravity acted at a distance. We cannot explain why brute regularities exist between consciousness and the physical world.

5.6 The Computational Discovery of Theories of Consciousness

> [...] the era of simple mathematics effectively modelling parts of the world is drawing to a close. It is possible that new areas of investigation will lend themselves to simple models, but the evidence is that within existing areas of investigation, the domain of simple models has been extensively mined to the point where the rewards are slim.
>
> Paul Humphreys, *Extending Ourselves*[27]

Traditional science is based on the idea that *people* identify regularities in the physical world. It is a working assumption that physical regularities are simple enough to be found by humans. When a human finds a regularity s/he might use a novel property to explain it. A mathematical description of the regularity can be experimentally tested. Humans find this satisfying—they like solving puzzles. But it is not necessarily the most effective approach.

Humans are biased and stupid. They have small working memories and little imagination. They cannot process large data sets. These limitations will prove fatal to consciousness science if there are complex relationships between c-descriptions and p-descriptions.[28]

We have little or no idea about the spatiotemporal structures that form CC sets. The mathematical relationships between c-descriptions and p-descriptions are unknown. They might be simple—a few differential equations. Or the mathematical complexity of these regularities could be way out of reach of human capabilities. Macro-scale laws of the brain could extend to thousands of pages of differential equations.

We should drop the *assumption* that there are simple relationships between p-descriptions and c-descriptions. We have no reason to believe that this is the case. If we persist with the assumption that there are simple relationships, we could spend large amounts of time and money on a fruitless quest for something that does not exist. It is better to assume that the relationships between c-descriptions and p-descriptions are *potentially* complex, and develop a methodology that can identify simple and complex relationships (or prove that no relationships exist).

We could use computers to identify the relationships between p-descriptions and c-descriptions. This would require a large amount of data, spanning multiple levels of the brain. C-descriptions and p-descriptions would have to be recorded for many different conscious states. This data could be gathered by human scientists. Or robots could capture it automatically.[29]

Machine-learning techniques could be applied to this data. The patterns that were found could be used to make predictions, which could be tested in further experiments. C-theories could be automatically tested against new data as it came in.

C-theories that are discovered by computers should be expressed in a format that can be read by both humans and machines (they should not be stored as weights in a complex neural network). This would enable them to be partly viewed and verified by humans—but there would be no expectation that an individual human scientist could check or comprehend them in their entirety. Sets of differential equations would be a good choice of output format—there is a long tradition of using differential equations to describe complex relationships in the physical world. Or perhaps we could use graph theory to describe the relationships between c-descriptions and p-descriptions.[30]

This computational approach to the science of consciousness could identify simple relationships between c-descriptions and p-descriptions. It could find regularities that are too complex to be identified by humans. Or it could prove that no simple or complex laws exist in the current data.

This approach could be prototyped on a simulated human brain.[31] This would generate reports 'about consciousness' that are similar to c-reports and could be converted into c-descriptions.[32] The computer could search for relationships between these c-descriptions and different aspects of the neural model. It could control the simulation (rewinding it, rewiring it, changing its parameters) to robustly test its hypotheses.[33] The relationships between c-descriptions and p-descriptions that were identified by this method could be tested on platinum standard systems.[34]

5.7 Summary

This chapter has described how we measure the physical world. Physical measurements have to be expressed in a formal way (a p-description), so that we can use our knowledge about CC sets in humans to make inferences about the consciousness of non-biological systems.

Pilot studies could identify the CC sets that are linked to individual conscious states. These must use natural experimental methods, which preserve our ability to measure consciousness in platinum standard systems.

In the longer term we need to develop mathematical c-theories that map between p-descriptions and c-descriptions (see Figure 5.2). These c-theories must conform to the constraints on CC sets (C1-C4).

Humans might be incapable of discovering complex mathematical relationships between p-descriptions and c-descriptions. To avoid this potential problem, computers should be used to discover c-theories.

6. Physical Theories of Consciousness

6.1 Physical C-Theories

The physical world contains elementary wave-particles (quarks, leptons, bosons). These are arranged into structures at different spatial scales. Quarks and electrons form atoms. Atoms are the constituent parts of molecules, which are the constituent parts of neurons, blood and bone. A structure at one level of the physical world will be referred to as a material:

> **D11**. A *material* is an arrangement of elementary wave-particles at a particular spatial scale.

The constituent parts of materials are formed from other materials — carpets are made from nylon, which is made from molecules, and so on. Some of a material's properties are not attributable to its constituent parts: water is wet; the electrons, protons and neutrons in water are not wet. Spatiotemporal patterns occur in materials (tartan, waves, etc.).

Physical c-theories are defined as follows:

> **D12**. A *physical c-theory* links consciousness to spatiotemporal patterns in materials. Physical CC sets consist of one or more patterns and the materials in which these patterns occur.[1]

Physical c-theories map p-descriptions of patterns in materials onto c-descriptions of conscious states. They also map c-descriptions of conscious states onto p-descriptions of patterns in materials.

In a physical c-theory the materials are essential members of the CC sets: each pattern has to occur in a particular material. A physical c-theory that links a conscious state to an electromagnetic wave pattern would not attribute consciousness to a pile of beer cans that happened to instantiate the same pattern.

Physical c-theories fit in neatly with the standard sciences (physics, chemistry, biology, geology), which identify patterns in particular physical things (planets, molecules, proteins, glaciers). They have the same level of objectivity as these other sciences (C1).

6.2 Potential Physical CC Sets in the Brain

The normally functioning adult human brain is our only platinum standard system (A4). So the science of consciousness can only look for physical CC sets that contain one or more of the materials that are present in the human brain and one or more of the spatiotemporal patterns that occur in these materials.

Some of the brain's properties depend on other parts of the physical world. For example, brains reflect electromagnetic waves and every neuron is a particular distance from the North Pole. These are not intrinsic properties, so they are not potential members of CC sets (C2).[2]

Physical CC sets can exchange physically conserved quantities, so they can e-cause c-reports during consciousness experiments (C4). It is not necessary that everything in a CC set e-causes c-reports, but the set as a whole must be capable of this. The changes in the balance of oxygenated and de-oxygenated blood that are measured by fMRI cannot e-cause c-reports because they peak several seconds after the c-report. Blood flow patterns that occur on an appropriate time scale are potential members of CC sets.[3]

The materials that could be members of CC sets include neurons, glia, blood, cerebrospinal fluid, electromagnetic waves, quantum states and novel materials.[4] With the possible exception of novel materials (see Section 6.3), a physical CC set cannot solely consist of materials, which are typically present when the brain is unconscious (C3). Some of the materials in a physical CC set must contain patterns that only occur when the brain is conscious.

The patterns in physical CC sets could be computational structures, such as a global workspace,[5] or patterns in the functional or effective connectivity between neurons.[6] I have suggested elsewhere that the neural patterns caused by sensory input could be linked to conscious sensations, and that a combination of sensory and sensorimotor patterns might be linked to our conscious perception of a three-dimensional world.[7] We could also use Tononi's information integration algorithms to identify patterns in materials that might be linked to consciousness.[8]

Some examples of physical CC sets:

- {neuron firing pattern p_3}
- {neuron firing pattern p_3, electromagnetic wave pattern p_4}
- {quantum pattern p_5}
- {neuron firing pattern p_3, haemoglobin}

In the last example the simple presence of haemoglobin is a member of the CC set. It does not matter which pattern occurs in the haemoglobin: a conscious state would only occur when p_3 is present in neurons surrounded by blood.

It is essential that the members of a physical CC set can be precisely and unambiguously described. Mathematical c-theories work with formally structured p-descriptions—they cannot convert natural language descriptions of the physical world into c-descriptions. It is easy to construct p-descriptions of elementary wave-particles, atoms and molecules. It is much harder to p-describe biological materials, such as neurons (see Section 5.1).[9]

6.3 Novel Materials?

It has been suggested that consciousness could be linked to unknown materials, such as a novel wave-particle.[10] The novel material could contain patterns that are linked to conscious states. Or it could be a passive member of CC sets that include patterns in other materials. It is conceivable that each conscious state is linked to a different novel material.

Novel materials must have e-causal powers. Novel materials without e-causal powers could not e-cause c-reports (C4) and they could not be detected with scientific instruments. So we could not verify their existence or use them to infer the presence of consciousness. This type of material should be sliced off with Ockham's razor.

Up to this point it has not been necessary to posit novel materials to explain the brain's operation.[11] Most scientists believe that the known physical properties of the brain can account for the firing patterns that send spikes to the larynx and lead to c-reports. Novel materials might be needed if we observed brain events that did not have an identifiable cause—this might lead us to hypothesize new wave-particles. But no such cases have come to light.

The most plausible novel material is something with weak e-causal powers that plays a minor role in the e-causation of c-reports. Such a consciousness force or particle might be detectable by special instruments, but it would be invisible to our current technology. There is no pressing need for a consciousness force or particle, but we might believe in it if it was a necessary consequence of a c-theory that had been thoroughly tested in other ways.[12]

6.4 Simplifying Assumptions about Physical C-Theories

Experiments on physical c-theories have to demonstrate that some patterns in some materials are linked to consciousness and other patterns in other materials are not. For example, a physical c-theory might claim that some neuron activity patterns form CC sets. To prove this we would need to show that consciousness is correlated with the proposed neuron activity patterns independently of glia patterns, electromagnetic wave patterns, other neuron activity patterns, and so on.

The best way to prove that consciousness is linked to patterns in particular materials is to carry out studies that test all combinations of materials (see Section 5.3). However, the link between consciousness and particular materials cannot be fully tested in natural experiments, because the brain does not naturally change into different materials, such as silicon (see Section 5.4). So we are unlikely to be able to identify the minimal sets of spatiotemporal structures that form physical CC sets. For example, we will be unable to experimentally distinguish between these potential CC sets:

- {neuron firing pattern p_3}
- {neuron firing pattern p_3, haemoglobin}
- {neuron firing pattern p_3, haemoglobin, cerebrospinal fluid}

Some potential CC sets can be eliminated by assuming that passive materials are not members of CC sets. For example, we can assume that the simple presence of haemoglobin is not linked to consciousness. This assumption should only be made when natural experiments cannot prove the link between consciousness and the simple presence of a material:

> **A7.** *CC sets do not contain passive materials.* If the link between consciousness and the simple presence of a material cannot be demonstrated in a natural experiment, then this material can be excluded from potential CC sets.

Passive materials are only passive relative to consciousness—the materials could contain patterns that are not correlated with consciousness.

We can also assume that constant patterns are not members of CC sets:

> **A8.** *CC sets do not contain patterns that are present when the system is conscious and unconscious.* If the link between consciousness and a constant pattern cannot be demonstrated in a natural experiment, then this pattern can be excluded from potential CC sets.

This assumption only applies to constant patterns that occur in the same materials as the patterns that are linked to consciousness. Constant patterns that occur in other materials can be excluded using assumption A7.

Suppose a conscious brain has neuron activity patterns p_6 and p_7, and the unconscious brain has neuron activity patterns p_6 and p_8. If a natural experiment cannot demonstrate that p_6 is correlated with consciousness, then it can be excluded from the CC set using A8. The CC set would just consist of neuron activity pattern p_7.

There are strong connections between the brain's materials. When a neuron fires, there are chemical changes, fluctuations in electromagnetic fields and an altered balance between oxygenated and de-oxygenated blood. These changes are partially correlated with each other, but only one of them might be linked to consciousness.

Some of these partially correlated changes can be excluded from potential CC sets on the basis of their timing relationship with c-reports. It is more difficult to identify the correlation between consciousness and patterns that occur simultaneously. This could be done by replacing the brain's materials, but there is little scope for this in natural experiments. To address this problem we can exclude weakly correlated patterns from potential CC sets:

> **A9.** *CC sets do not contain partially correlated patterns.* When several different materials have the same spatiotemporal pattern, the material(s) in which the spatiotemporal pattern is strongest will be considered to be the potential member(s) of the CC set that is associated with the conscious state, unless the partially correlated patterns can be separated out in a natural experiment.[13]

Suppose a conscious brain has neuron activity pattern p_9 with strength 7, and p_9 occurs in the glia with strength 5 and in the electromagnetic waves with strength 8. p_9 is completely absent from the unconscious brain. A natural experiment cannot demonstrate that p_9 in neurons and glia should be excluded from the CC set, so the CC set could consist of p_9 in any combination of the three materials. We can set this possibility aside by making assumption A9. The CC set would just consist of electromagnetic wave pattern p_9.

A7-A9 enable us to develop more compact c-theories from ambiguous experimental data. They should not be rigidly adhered to because they are motivated by pragmatic considerations and go beyond the experimental evidence. CC sets might contain passive materials, constant patterns and partially correlated patterns.

6.5 Summary

A physical c-theory is a mathematical relationship between patterns in materials (captured in p-descriptions) and formal descriptions of conscious states (c-descriptions). Physical c-theories conform to the constraints and fit in well with mainstream scientific methodology.

A large number of materials and patterns are potential members of physical CC sets. We are unlikely to be able to completely separate them out in natural experiments, but we can reduce the number of potential CC sets by making assumptions A7-A9, which exclude passive materials, constant patterns and partially correlated patterns from CC sets.

7. Information Theories of Consciousness

[…] to the extent that a mechanism is capable of generating integrated information, no matter whether it is organic or not, whether it is built of neurons or of silicon chips, and independent of its ability to report, it will have consciousness.

Giulio Tononi, Consciousness as
Integrated Information: A Provisional Manifesto[1]

Information is notorious for coming in many forms and having many meanings. It can be associated with several explanations, depending on the perspective adopted and the requirements and desiderata one has in mind.

Luciano Floridi, Information: A Very Short Introduction[2]

7.1 What Is Information?

In this 'information age' people see information everywhere. Some say that we are living in a simulation or a digital universe; others claim that information patterns *are* consciousness.

I open up your head and rummage around inside. I feel bones, blood and tapeworm cysts. Through the microscope I observe neurons, glia and bacteria. I cannot see *information* anywhere. I cannot detect it using scientific instruments. There is just soggy oozing physical stuff.

Computers are information processors. There must be information inside a computer. I open up a computer and rummage around inside. I feel silicon chips, copper circuits, dust and two dead flies. Just more physical stuff—no information anywhere.

I flick through the computer manual. It states that information is stored in the memory units of the computer (the DRAM storage cells).

I switch on the computer and examine the DRAM storage cells. They contain electrons. When I measure the voltages I obtain the following values: 0.7, 0.8, 1.1, 1.0, 0.2, 0.7, 0.9, 0.1, 0.0, 1.5, 1.4, 0.5, 0.1, 1.5, 0.7, 0.8, 0.3, 1.2, 1.3, 0.0, 0.4, 0.9, 0.7 and 0.6. These voltages change all the time as the computer operates. If an engineer looked these voltages, s/he would note that the DRAM is operating in its specified range.

I apply a threshold of 0.75 V to the voltages, and interpret voltages above the threshold as 1 and voltages below the threshold as 0. This yields 011100100110010101100100. This means something to me—it is a string of 1s and 0s. A computer scientist exclaims, 'Ah, binary, that's 726564 in hexadecimal.' A child interprets it as an adder.

I group the 1s and 0s into three 8-bit binary numbers: 01110010, 01100101 and 01100100. These correspond to the decimal numbers 114, 101 and 100. I map the decimal numbers onto letters using the standard ASCII codes (114='r'; 101='e'; 100='d'). This yields 'red'. It is a word in the English language (the colour of apples; the colour of blood). 'red' does not mean much to people who do not speak English—it is just a string of letters, similar to 'nob'.

Initially the computer was an invisible physical object. I did not attribute any properties to it. It was something beyond my bubble of experience that did not exist for me. Its physical states were not 1s and 0s; they were not letters or numbers; they were not even voltages.[3]

Voltages, binary numbers and 'red' are information patterns that appear when we measure a system's states and interpret the measurements in different ways. This combination of measurement and interpretation will be referred to as an *interface*, which specifies:

- The material that holds the information. In a computer the information could be in the DRAM, CPU, etc.
- The type of information. Is it binary, decimal, drawn from the set of letters, and so on?
- How information of the appropriate type can be read from spatiotemporal patterns in the material. In the computer example I specified how the DRAM voltages could be measured and converted into binary numbers and letters.

Interfaces enable us to extract information from the invisible physical world. They can be applied in sequence to extract different kinds of information. There is no information without an interface.[4]

An infinite number of different interfaces can be applied to a physical system. Instead of a threshold of 0.75 V I could have used a threshold of 0.55 V. This would have yielded 111101100110011101100111. I can group these 1s and 0s into four 6-bit binary numbers: 111101, 100110, 011101 and 100111, which correspond to the decimal numbers 61, 38, 29 and 39. I can use a different mapping of numbers onto letters (for example, 61='b', 38='l', 29='u' and 39='e'). This interface extracts 'blue' from the voltages in the computer's memory.

'Red' and 'blue' appear when I apply different interfaces to the DRAM voltages. There is no correct answer about which sequence of letters is *really* in the computer's memory. Different interfaces produce different sets of information.

Once I have selected an interface, the information is determined by the physical system. If I interpret the DRAM voltages using a threshold of 0.75 V, 8-bit numbers and standard ASCII codes, I inevitably end up with the word 'red' — I cannot change the fact that the application of this interface to this system results in the word 'red'. While information can only appear through a subjectively chosen interface, it is fixed by the physical system once the interface has been selected — it is objectively present on the basis of this interface.

Custom interfaces can be designed to read most and possibly all information patterns from a physical system in a particular state. I can extract the text of *Madame Bovary* from the lines on my wife's face.[5] Think of a four letter word — I can extract it from the DRAM voltages by changing the number-to-letter mappings. Time-indexed interfaces might be required to extract complex information from simple systems[6] and to extract sequences of information patterns from sequences of physical states.[7]

Some people distinguish data from information. They define data as the differences that are extracted from a physical system using an interface. These differences become information when they are well-formed and meaningful.[8] The problem with this distinction is that any measured set of differences is meaningful to some extent: Voltages are meaningful to engineers; binary numbers are meaningful to computer scientists; letters are meaningful to literate people. The only differences that are completely without meaning cannot be accessed by us because they are part of the invisible physical world. This leaves us with the notion that information might be *well-formed* data. But we do not need a data/information distinction to capture the difference between well-formed and badly-formed data.

Shannon's work on the transmission of information has led some people to interpret information as the reduction of uncertainty.[9] Consider my snake, Sam. Sam is dead. Sam is not Lazarus: he will not rise—he will always be dead. You do not need to tell me that Sam is dead because I know that he is dead and this is not going to change. I do not gain any information when you send me a message stating that Sam is dead. Now consider a coin that can be in two states (heads and tails). I gain information (I reduce uncertainty) if you tell me that it is tails because I can only guess this with 50% accuracy. Now consider a six-sided dice. You roll the dice and it shows a two. I can only guess that it is showing two with 17% accuracy, so a message informing me that it is two considerably reduces my uncertainty about it. The more a message reduces my uncertainty about the state of a system, the greater the information content of that message. Shannon used this interpretation of information to develop his measure of information entropy.

This interpretation of information is a useful way of *quantifying* the amount of information in a system. But it is not an adequate definition of information. Before we can talk about the reduction of uncertainty of our knowledge about a system, we need an interface that defines the information states that are available in the system. We can only reduce uncertainty about the state of a coin once we have an interface that converts the physical coin into two possible outcomes, 'heads' and 'tails'. Once a system's information states have been defined, it is possible to measure its information entropy and state how rapidly its information can be passed over a channel.

7.2 Information C-Theories

Information c-theories are defined as follows:

D13. An *information c-theory* links consciousness to spatiotemporal information patterns. Information CC sets only contain information patterns, which can occur in any material.

Suppose we discover a neuron firing pattern, p_{10}, that is correlated with conscious state c_4. We could apply an interface, i_1, to this pattern to extract an information pattern, ip_1. An information c-theory would claim that c_4 is correlated with ip_1. This c-theory would predict that c_4 would be present if ip_1 was extracted from a pile of stones or from a set of traffic lights (see Figure 7.1).

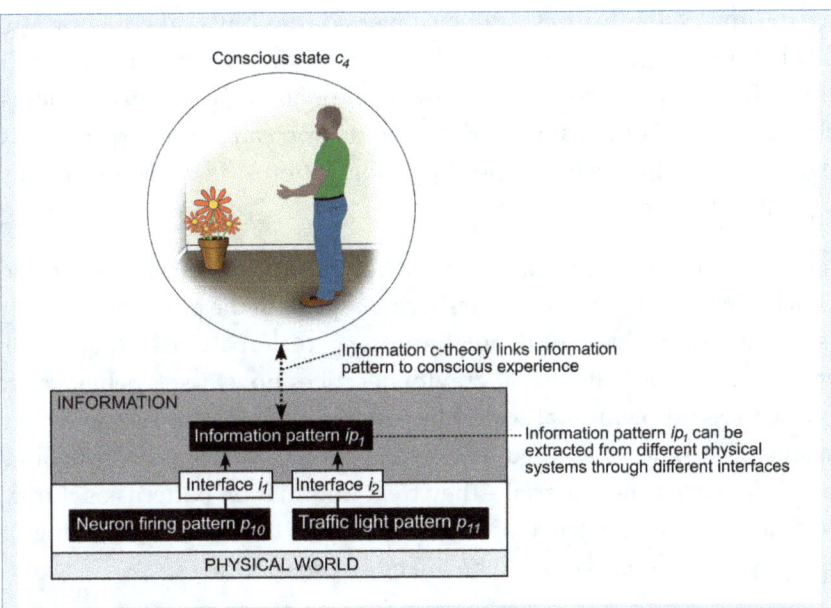

Figure 7.1. Information c-theory. An experiment demonstrates that conscious state c_4 is correlated with neuron firing pattern p_{10}. Interface i_1 converts neuron firing pattern p_{10} into information pattern ip_1. An information c-theory would claim that it is the information pattern, ip_1, that is linked to c_4, not the neuron firing pattern p_{10}. Information pattern ip_1 can also be extracted from traffic lights through interface i_2. The information c-theory would claim that ip_1 is linked to c_4 regardless of whether it has been extracted from a neuron firing pattern or a set of traffic lights. Image © David Gamez, CC BY 4.0.

Tononi has developed an impressive information c-theory. His algorithm analyzes the information patterns in a system, and outputs the parts that are linked to conscious states, the level of consciousness and a high-dimensional mathematical structure that is intended to correspond to the contents of consciousness. Preliminary experiments have been carried out to test this c-theory.[10]

Physical c-theories use interfaces to gather information about the physical world. The interface acts as a window onto the materials, and physical c-theories link patterns in these materials to consciousness. In an information c-theory the information pattern that is extracted through an interface is not a measurement of something else—it is linked to consciousness independently of the interface or the material in which it occurs.

Information c-theories can be converted into physical c-theories by adding material(s) to the CC sets.[11] Physical c-theories can be converted into information c-theories by removing the material(s). The experiments that support Tononi's information c-theory can be interpreted as evidence for a link between neuron firing patterns (identified using his algorithm) and consciousness.

Information c-theories are a radical departure from standard scientific practice. Scientific laws apply to specific aspects of the physical world. It is not the pattern that counts, but the presence of the pattern in a particular material. Newton's theory of gravity describes how *masses* behave on a particular spatiotemporal scale. His equations would produce incorrect results if they were applied to electric charges. Information c-theories break free from the material—they treat information patterns as if they had an objective existence of their own—as if they were something *in* the physical world that could be linked to consciousness.

7.3 The Subjectivity of Information

A brain is in state s_1; conscious state c_5 is present. My laboratory carries out a pilot study to identify the information pattern that is correlated with c_5. Tony chooses one interface and claims that the resulting information pattern, ip_2, is correlated with c_5. George chooses a different interface and claims that the resulting information pattern, ip_3, is correlated with c_5. Which information pattern is correlated with c_5—ip_2, ip_3 or both?

Tony is my pal. George broke my microscope. I want to accept Tony's claim that ip_2 is correlated with c_5. I want to reject the information pattern gathered by clumsy George. But the selection of ip_2 would be an arbitrary subjective choice. If I want to conform to constraint C1, I have to accept that any and potentially all of the information patterns that can be extracted from the brain in state s_1 are potentially correlated with c_5. To avoid subjectivity my pilot study will have to measure them all. This is impossible because there is an infinite number of them.[12]

Suppose I use all possible interfaces to measure all of the information patterns that can be extracted from the brain in state s_1. I now need to identify the ones that are *correlated* with c_5. Which of these information patterns are *not* present in the unconscious brain?

My pilot study will have to use all possible interfaces to measure all possible information patterns in the unconscious brain. I can then compare these infinite sets to find the information patterns that are only present in the conscious brain. These are the members of the information CC set that is correlated with c_5. The practical impossibility of this task suggests that a subjective choice of interface cannot be avoided in real world experiments on information c-theories.

These practical difficulties are irrelevant if interfaces can be *custom designed* to extract arbitrary information patterns from the brain (see Section 7.1). This would enable *any* information pattern to be read from the unconscious brain, including the information patterns that were extracted from the conscious brain in the first stage of the experiment. Custom designed interfaces that can extract arbitrary information patterns would break constraint C3. CC sets cannot consist of information patterns if all of the conscious brain's information patterns can be extracted from the unconscious brain.[13]

7.4 E-Causal Powers of Information

I define an interface that interprets voltages in a computer's memory as 1 if they are above 0.75 V, and as 0 if they are below 0.75 V. The information changes as the voltages change. This interface makes no difference to the patterns of e-causation in the computer—with and without the interface the computer moves through the same sequence of physical states.[14]

I alter the interface and specify that voltages above 0.8 V should be interpreted as 0, and voltages below 0.8 V should be interpreted as 1. Now the information patterns are completely different, but the computer continues to move through the same sequence of physical states. It does not matter which interface I apply to the computer: its e-causal exchanges and sequence of physical states remain the same. The information does not e-cause or constrain the behaviour of the physical system. This suggests that information cannot e-cause c-reports, so information patterns cannot be sole members of CC sets (C4).[15]

7.5 Is Information Intrinsic?

Information appears when an interface, defined by an observer, is applied to a physical system. Information patterns that depend on an external interface cannot be intrinsic properties.

It is conceivable that the interface could be inside the system, so that one part reads information from another.[16] In this case the information might be an intrinsic property of the system as a whole. There are problems with this proposal. For example, the location of the information patterns in the system would be ambiguous, and information c-theorists have not proposed how we can measure this type of 'intrinsic' information without applying an external interface to the system.

7.6 Separating Information from Material

Suppose we identify a neuron firing pattern that is correlated with a conscious state. There are two interpretations of this result:

- A pattern of information is linked to the conscious state (information c-theory).
- A pattern in a material (neurons) is linked to the conscious state (physical c-theory).

We want an experiment that can decide between these two claims. This would show that an information pattern is correlated with consciousness (information c-theory), or that the pattern is only correlated with consciousness when it occurs in biological neurons (physical c-theory).

The best way of deciding between these claims would be to change the brain's materials while preserving its information patterns. If it had the same conscious state when its neurons were replaced with silicon, then the information pattern might be the sole member of the CC set. But if we exchange a person's neurons for silicon, we cannot be confident that their c-reports are functionally connected to their conscious states. We will have lost our ability to measure consciousness (see Section 5.4).

We have to use natural experiments to decide between the two claims. We could monitor the system and hope that the pattern moves between materials during its normal behaviour. Suppose the subject has conscious state c_6 when there is information pattern ip_4 in the neurons and nowhere else in the brain. At a later point in time the subject has c_6 when ip_4 is in the glia and nowhere else in the brain. We would conclude that c_6 is correlated with the information pattern, and that the material has no effect on consciousness.[17]

We have no reason to believe that information patterns move between materials during natural experiments on the brain. If we cannot observe this, it will be impossible to experimentally distinguish between physical and information c-theories.

7.7 Summary

Information appears when interfaces are applied to the physical world. Interfaces specify how information of a particular type can be extracted from a particular material. Information does not exist *in* the physical world—it is partly determined by the interface and partly determined by the physical world. Information c-theories claim that information patterns are linked to consciousness independently of the material in which they occur.

Information patterns cannot be correlated with consciousness because they can be read from both the conscious and unconscious brain using custom-designed interfaces (C3). Information patterns are subjective and incapable of e-causing c-reports (C1, C4). It is unlikely that evidence in favour of them can be obtained through natural experiments. Until these problems have been resolved information c-theories should be set aside or interpreted as physical c-theories.[18]

8. Computation Theories of Consciousness

> Useful computation is in the eye of the beholder. [...] It requires an underlying system of whose autonomous dynamics we have a predictive model. [...] To solve a problem computational we need to map the problem to be solved onto the underlying behavior of the system and hence produce a starting state from which the autonomous dynamics of the system will produce a solution.
>
> Robert Kentridge, Symbols, Neurons, Soap-Bubbles and the Computation Underlying Cognition[1]

8.1 Calculators, Special-Purpose Computers and General-Purpose Computers

We can solve many problems in our heads. But it is often easier to use the physical world to solve problems. Suppose you want to calculate 10+4+18+2. Put ten stones in a box, then four, and so on. Count the number of stones in the box to get the result. This is a simple *calculator*. The abacus, slide rule and Pascaline are more sophisticated calculators. When you enter a problem into a calculator the solution is immediately displayed. As each stone is put in the box you can immediately read off the total number of stones—there is no waiting while the system 'computes'.

Special-purpose computers take time to solve problems. We enter the problem by modifying the system's state (turning knobs, pressing keys, etc.). We set the special-purpose computer running and it transforms the starting state into the final state. We read off the result from the final state.

Special-purpose computers can only solve a limited number of closely related problems. A special-purpose computer that uses water to

© David Gamez, CC BY 4.0

model an economy[2] cannot process words or simulate an aircraft wing. Turing machines are special-purpose computers.

Soap bubble computers can find a short path between multiple points (see Figure 8.1). Other special-purpose computers are Turing's Bombe, which was used to crack German encryption during the Second World War, and Babbage's difference engine—a mechanical system that uses gears, cams, rods, levers and springs to compute polynomial functions.

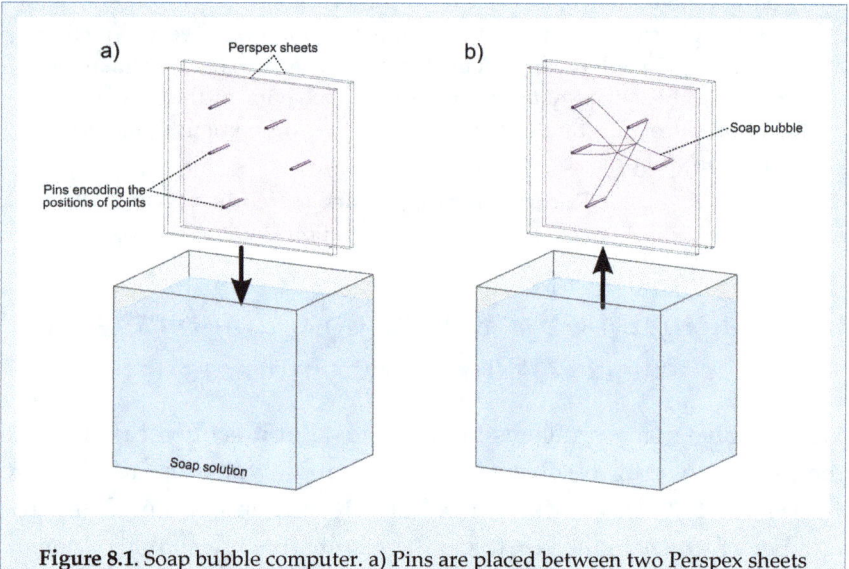

Figure 8.1. Soap bubble computer. a) Pins are placed between two Perspex sheets to indicate the positions of some points. b) The sheets are dipped in soap solution and the resulting bubble contracts to minimize its surface tension, producing a short path between the points. Image © David Gamez, CC BY 4.0.[3]

General-purpose computers run programs. Programs turn a general-purpose computer into a special-purpose computer for a limited period of time. The operator specifies the program by connecting wires, punching cards, typing code into a terminal, and so on. The program puts the computer into a starting state. Then the computer's components interact according to the laws of physics, and the result is read off from the computer's finishing state. Universal Turing machines are general-purpose computers.

Babbage's Analytical Engine is a design for a general-purpose computer that uses gears, cams, rods, levers and springs to run programs. The ENIAC was the first general-purpose computer to be built. By manipulating switches and cables it could be rewired into a special-purpose computer that corresponded to a desired program. The innovation of the ENIAC was that it was designed to be rewired easily—previous computers could only run a limited range of programs without being heavily modified (by adding components, re-soldering circuits, etc.). The Manchester Baby was the first general-purpose computer that could 'rewire' itself electronically from a program stored in a cathode ray tube.

Modern digital computers are general-purpose computers. Their programs are typically written in a high level language, such as C++. Another program, called a compiler, converts this human-readable list of instructions into binary machine code, which is stored as a set of voltages in the computer's memory. When the program runs, the voltages in the computer's memory interact with the CPU and other components. This causes the computer to change states in a sequence determined by the program. When the program finishes the result is read back from the computer's memory.

Special-purpose computers can execute the same computations as general-purpose computers (the same computations can be executed by Turing machines and universal Turing machines). The key difference is that general-purpose computers can be dynamically reconfigured to run different programs. Special-purpose computers have to be custom-built to run a particular program.

When a general-purpose computer runs a program it is executing the computations that are specified in the program—its *potential* ability to execute other programs does not affect its *current* computations. General-purpose computers operate as special-purpose computers when they are running a program.[4]

The brain can operate as a general-purpose computer. I can manually execute a program, using pen and paper to keep track of the variables.[5] When a brain is not manually executing programs, it is not operating as a general-purpose computer. In ordinary life the brain works as a special-purpose computer: its state transitions are determined by complex biological structures that are mostly hardwired.

8.2 Computation C-Theories

Computers are artificially intelligent—they can chit-chat, fly planes and play games. A computer that was simulating your brain might behave just like you. In a few short years computers will conquer the world and grind our weak human flesh into fertilizer.

Computation c-theories are defined as follows:

D14. A *computation c-theory* links consciousness to the execution of computations. Computation CC sets only contain computations, which can be executed by many different types of computer.[6]

Computation c-theories are motivated by the observation that computers can do the same things as brains. So it is hypothesized that brains work in a similar way to computers. Computation c-theories are encouraged by the fact that a program can run on many different types of machine. This suggests that computations could be objectively present in many different materials.

Computation c-theories claim that computation CC sets are linked to consciousness independently of the system that is executing them. So a computation CC set would be linked to consciousness if it was executed manually, if it ran on the cogs of Babbage's Analytical Engine or if it was executed on the voltages of a modern digital computer.

The brain is conscious when it is operating as a special-purpose computer. So any computations that might be linked to consciousness can be executed on special-purpose computers. The convenient features of general-purpose computers are irrelevant to computation c-theories.[7]

8.3 The Subjectivity of Computing

My mate Crystal has asked me to father her children. She wants to be impregnated when the planets are aligned. That way, her children will inherit my good looks *and* become great warriors. To calculate the date I need a model of the solar system.

The solar system can be simulated on a digital computer. I write a program that uses Newton's equations to manipulate numerical representations of the masses, positions and velocities of the sun and planets. A compiler converts the program into machine code. When I load the compiled program it becomes a pattern of voltages in the computer's memory. When I run the program the components in the computer interact according to the laws of physics. After a few seconds the program terminates and I interpret the pattern of output voltages as the date of planetary alignment.

The solar system can be modelled by an orrery, which uses a clockwork mechanism to move metal balls representing the planets around a ball representing the sun. To calculate the date of planetary alignment, I put the spheres into the planets' current positions, set the mechanism running and read off the date when they align.[8]

I can increase the accuracy of my orrery by discarding its clockwork mechanism and replacing its balls with metal spheres that have the same masses as the sun and planets. To set up this gravity-powered orrery I put the spheres into positions that correspond to the sun and planets and give them the same velocity. I allow them to rotate under the influence of gravity and note when they align.

It makes no difference to my model if it uses metal spheres or the actual sun and planets. The latter approach requires less effort. The planets are already in their starting state and moving under the influence of gravity. I allow them to rotate and read off the date when they align.

All of these systems are computing the same global function: the date when the planets align. If *any* of these systems have computational properties (that are potentially linked to consciousness), then *all* of these systems have computational properties (that are potentially linked to consciousness). Computers can be made from silicon, metal or planets—it does not matter if they are powered by clockwork or gravity.

Prior to my intervention the solar system was not a special-purpose computer: it was not *calculating* the paths of the planets. It was just a group of massy bodies whose state changes were dictated by the laws of physics. The solar system became a computer when I used it to calculate the paths of the planets. But its new status did not affect its material properties or behaviour—it continued to follow the laws of physics in exactly the same way as before. This suggests that computers are subjective interpretations of the physical world—part of the physical world *becomes* a computer when I use it to solve problems.

When I use my iPad to bang in a nail it can be useful to describe it as a hammer. Some of its properties can be understood by comparing it with other hammers. But my iPad does not contain 'hammutations'—I do not need to invoke 'hammutations' to explain how I can bang in a nail with my iPad.

When we use part of the physical world to add numbers, it can be useful to describe it as a calculator. Some of its properties can be understood by comparing it with other calculators. But the physical world does not *contain* calculations. Stones and a box do not contain calculations—I use them to add numbers.

When we use the state changes of physical objects to solve problems it can be useful to describe them as computers. Some of their properties can be understood by comparing them with other computers (a digital computer is faster and more flexible than an orrery). But I do not need to invoke the objective presence of computations in the physical world to explain how I can use digital computers, orreries and solar systems to compute the dates of planetary alignment.

The key difference between digital computers, orreries and solar systems is the extent to which they have been engineered to facilitate our use of them as computers. The solar system has not been engineered at all—it is difficult to set up in a desired starting state and it works on the same time scale as the system it is modelling. Clockwork orreries can be set up easily and work faster, but they can only model one type of system. Digital computers can run many different programs and they typically operate much faster than the systems they are modelling.

If computing is a use that we make of physical objects, then computations cannot be members of CC sets (C1). My consciousness does not appear when someone uses my brain as a computer.

8.4 Information Processing in Computers

Computers are often described as information processing technology. Chapter 7 explained how we use interfaces to extract information from the physical world. Interfaces can also be used to store information in the physical world.[9] I can write and read the same number, the same information, to and from many different physical systems.

Information that is stored in the physical world can be altered by changes in the physical world. I write Felicity's number on a piece of paper and store it as a sequence of pits on a compact disc. A tea stain blurs '7' into '8'. A scratch on the disc scrambles Felicity's number.

We use changes in the physical world to process information. We construct a system that changes in a systematic way. Then we modify part of the system to encode the information that we want to transform (this modification is determined by the interface). We allow the system to change state (to compute). When it has finished we use the same interface to read back the processed information from the system.[10]

We know that soap bubbles contract to minimize their surface area, but we do not know exactly *how* they will contract between a given set of pins — if we knew this, there would be no point in using a soap bubble computer to identify a short path. The encoding of positions using pins, the dipping in soap solution and the examination of the resulting bubble are worthwhile because they enable us to read back a short path solution that would be more complicated to obtain in other ways.

Digital computers are engineered to carry out fast and flexible information processing. We initialize a computer by creating voltage patterns in its DRAM that correspond to a program and initial data. The computer then moves through the sequence of physical states that is determined by the data and the program. Digital computers also include a physically implemented interface that uses carefully designed interactions between components in the screen, circuitry and chips to convert the DRAM voltages into graphical shapes painted in light, which we interpret as letters, numbers, etc.

The information processing that is carried out by a physical system depends on the interface that is used to read and write the information.

The soap bubble computer can be interpreted as processing information about the shortest roads between cities or about the optimal wiring of electronic components.

Suppose a computer's memory changes from 011100100110010 101100100 to 011100110111010101101110.[11] The information extracted through one interface (8-bit numbers, standard ASCII codes, 114='r', 101='e', 100='d', 115='s', 117='u', 110='n') changes from 'red' to 'sun'. Through a different interface (6-bit numbers, 28='r', 38='u', 21='i', 36='n', 55='a', 46='d'), the information changes from 'ruin' to 'raid'. There is no single correct or objective answer about the information processing that is being carried out by this computer. At most we can say that at least one interface exists that leads to the processing of 'red' into 'sun'.[12]

Information processing is not a unique attribute of computers or brains. Any system can be interpreted as an information processor. A digital computer does not process any *more* information than a tub of worms. But we can carry out more *useful* information processing with a digital computer.

Any system that is interpreted as an information processor inherits all of the problems with information that were highlighted in the previous chapter. Information is subjective (C1), it is not likely to be intrinsic (C2) and it does not have e-causal powers (C4). Most or all information sets can be read from the conscious and unconscious brain (C3). If computation is information processing, it cannot be a member of a CC set.

8.5 Digital Physics and Theories of Implementation

Keith is taking time out from his IT support work. He flops into a chair, pushes back his lank long hair, sparks up a joint and relaxes. Suddenly he has a vision of the universe as a giant computer.

Digital physicists claim that digital computation is a fundamental property of the universe.[13] Computation cannot be subjective if everything is computing all the time, regardless of whether we are using it to process information. This claim needs to be supported with a definition of what it means to implement a computation. Digital physicists cannot claim that everything is X without specifying the nature of X. When we have a theory of implementation, we can look for computational structures in the universe. If they are ubiquitous and play a fundamental physical role, then it can be claimed that the universe is a giant computer.

A theory of implementation is also required to test computation c-theories. Suppose we want to carry out a pilot study that looks for the computation CC set that is linked to a conscious state. We will need a theory of implementation that maps the brain's physical states onto computations. This will enable us to identify the computations that are executed in the conscious brain and not executed in the unconscious brain.[14]

Many theories of implementation have been put forward. None of them are convincing. Theories based on finite state automata lead to panpsychism.[15] Combinatorial state automata don't work.[16] Some theories of implementation are based on features of modern digital computers, such as string processing, which do not generalize easily to biological systems.[17] Many digital physicists favour cellular automata, but it is far from obvious whether cellular automata can provide a plausible interpretation of the physical world or the systems we call computers.[18]

Digital physics cannot rescue computation c-theories from subjectivity without a plausible theory of implementation. Computation c-theories cannot get off the ground without a theory of implementation that would enable them to be experimentally tested. We do not have a workable theory of implementation.

8.6 Summary

Computation c-theories claim that computations are sole members of CC sets — the architecture and material of the systems that are executing the computations are irrelevant. The convenient features of general-purpose computers are not necessary to computation c-theories. Any computation that is potentially linked to consciousness can be executed on a special-purpose computer.

If computation is a subjective use we make of the world, then computations cannot be members of CC sets (C1). If computers are information processors, computation c-theories will have the same problems as information c-theories (C1-C4). Digital physicists claim that digital computation is a fundamental property of the universe. But no one has developed a theory of implementation that convincingly supports digital physics or would enable us to identify computational correlates of consciousness in the brain. Until these problems have been resolved, computation c-theories should be set aside or interpreted as physical c-theories.[19]

9. Predictions and Deductions about Consciousness

9.1 Predictions about Consciousness

> I shall certainly admit a system as empirical or scientific only if it is capable of being tested by experience. These considerations suggest that not the verifiability but the falsifiability of a system is to be taken as a criterion of demarcation. In other words: I shall not require of a scientific system that it shall be capable of being singled out, once and for all, in a positive sense; but I shall require that its logical form shall be such that it can be singled out, by means of empirical tests, in a negative sense: it must be possible for an empirical system to be refuted by experience.
>
> Karl Popper, *The Logic of Scientific Discovery*[1]

Information and computation c-theories do not conform to constraints C1-C4. The rest of this book will focus on physical c-theories, which are based on the idea that patterns in one or more materials are linked to conscious states (D12).

Physical c-theories convert descriptions of physical states into descriptions of conscious states. This enables them to be tested in the following way:

1. Measure aspect of the physical world that is specified by the c-theory.
2. Convert measurement into p-description, pd_1.
3. Use mathematical c-theory to convert p-description into c-description, cd_1.
4. Obtain c-report from test subject.
5. Convert c-report into c-description, cd_2.
6. Compare cd_1 and cd_2. If they match, the c-theory passes the test for this physical state and this conscious state.

For example, we could measure the state of a person's brain and use a c-theory to generate a prediction about their consciousness. This is illustrated in Figure 9.1.

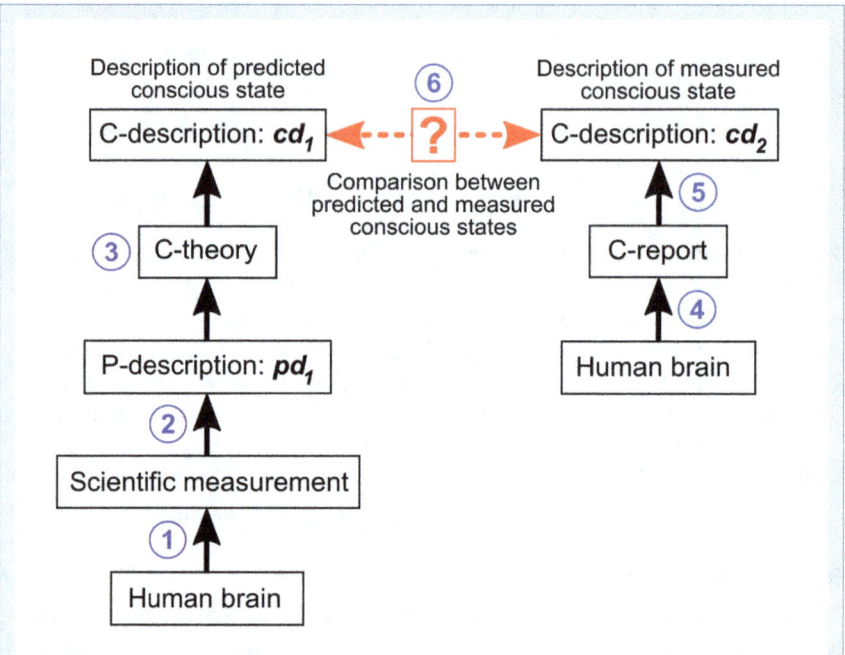

Figure 9.1. Testing a c-theory's prediction about a conscious state. 1) Scientific instruments measure the physical state of the brain. 2) Scientific measurements are converted into a formal p-description of the brain's physical state, pd_1. 3) C-theory converts p-description, pd_1, into a formal c-description of the brain's predicted conscious state, cd_1. 4) The human brain generates a c-report about its conscious state. 5) The c-report is converted into a formal description of the measured conscious state, cd_2. 6) The measured and predicted conscious states are compared. If they do not match, the c-theory should be revised or discarded. Image © David Gamez, CC BY 4.0.

The validation of the predicted consciousness (stages 4–6) could be carried out by the subject. First the predicted conscious state could be induced in the subject. Then the subject would compare the induced state of consciousness with their memory of their earlier conscious state (the state they had when their physical state was measured).[2]

Virtual reality could be used to induce predicted states of consciousness in the subject. The c-description of the predicted consciousness would be converted into a virtual reality file.³ This would be loaded into a virtual reality system and the user would decide whether their consciousness in the virtual reality system was similar to their earlier conscious state.⁴ We could also develop algorithms that convert c-descriptions of predicted conscious states into natural language. The subject could decide whether the natural language description corresponded to their memory of their earlier conscious state.

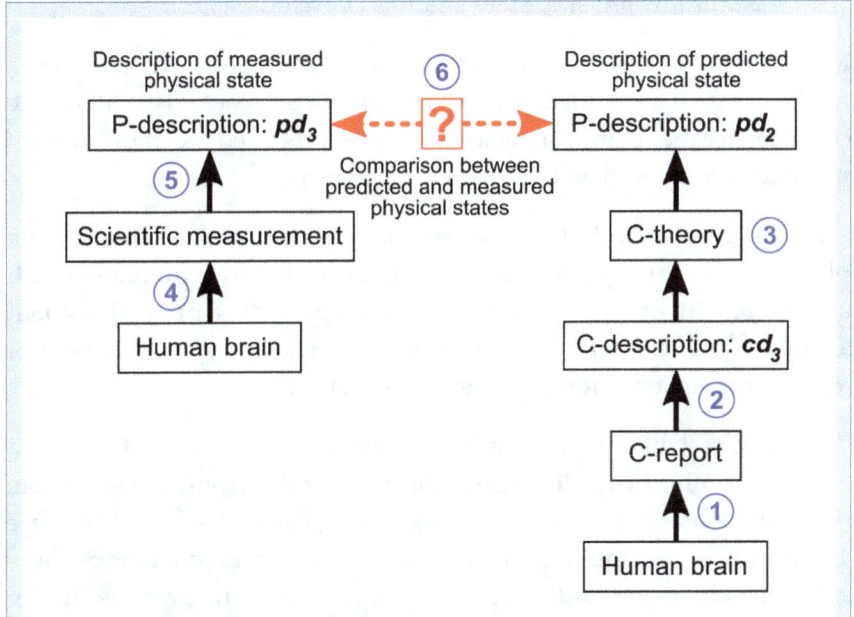

Figure 9.2. Testing a c-theory's prediction about a physical state. 1) The human brain generates a c-report about its conscious state. 2) The c-report is converted into a formal description of the measured conscious state, cd_3. 3) C-theory converts the c-description, cd_3, into a formal p-description of the brain's predicted physical state, pd_2. 4) Scientific instruments measure the physical state of the human brain. 5) Scientific measurements are converted into a formal p-description of the brain's physical state, pd_3. 6) The measured and predicted physical states are compared. If they do not match, the c-theory should be revised or discarded. Image © David Gamez, CC BY 4.0.

C-theories can convert descriptions of conscious states into descriptions of physical states. So they can also be tested in the following way (see Figure 9.2):⁵

1. Obtain c-report from test subject.

2. Convert c-report into c-description, cd_3.
3. Use mathematical c-theory to convert c-description into p-description, pd_2.
4. Measure aspect of the physical world that is specified by the c-theory.
5. Convert measurement into p-description, pd_3.
6. Compare pd_2 and pd_3. If they match, the c-theory passes the test for this physical state and this conscious state.

For example, a c-theory might predict that a conscious experience of a red rectangle is associated with a particular neuron activity pattern. We can measure the brain of a person who c-reports a red rectangle to see if the predicted neuron activity pattern is present.

C-theories can only be tested on platinum standard systems. On a platinum standard system we can compare a c-theory's prediction with a measurement of consciousness. Or we can generate a prediction about a physical state from a measurement of consciousness. This type of testable prediction is formally defined as follows:

D15. A testable *prediction* is a c-description that is generated from a p-description or a p-description that is generated from a c-description. Predictions can be checked by measuring consciousness or they are generated from measurements of consciousness. Predictions can only be generated or confirmed on platinum standard systems during experiments on consciousness. It is only under these conditions that consciousness can be measured using assumptions A1-A6.

Good c-theories generate many testable predictions. We believe c-theories to the extent that their predictions have been successfully tested. Different c-theories can make different testable predictions — we use this to experimentally discriminate between them.

The tests described in this section only check that a c-theory maps between conscious states and particular aspects of the physical world. They do not check that a c-theory is based on *minimal and complete* sets of spatiotemporal structures (D5). Suppose we have shown that a c-theory maps between neuron activity patterns and consciousness in normally functioning adult human brains. To test the theory fully we have to check that this mapping exists independently of the presence of other materials in the brain, such as electromagnetic waves, glia, haemoglobin and cerebrospinal fluid.

To prove that a c-theory is based on minimal and complete sets of spatiotemporal structures we need to vary the physical world systematically in the manner described in Section 5.3. Many variations of the physical world cannot be achieved with natural experiments. So it is extremely unlikely that the predictions of a c-theory can be fully tested.

9.2 Deductions about Consciousness

I put your head into a guillotine and chop it off. It falls into a basket. I watch your face. Your eyes move; your mouth opens and closes; your tongue twitches. These movements cease. I connect an EEG monitor to your brain. It is silent. After a couple of minutes a wave of activity occurs that fades away after twenty seconds.[6]

Your decapitated head might be associated with a bubble of experience long after it has been cut off. I cannot measure its consciousness because it is not a platinum standard system. But I can use a theory of consciousness that has been tested on platinum standard systems to make inferences about the consciousness of your decapitated head.

How does consciousness change during death? Which coma patients are conscious? When does consciousness emerge in the embryo or infant? How will my consciousness be affected by a brain operation? Can I copy my consciousness by simulating my brain on a computer? What are the bubbles of experience of bats, cephalopods and plants? Are robots conscious?

Suppose we converge on a c-theory that is commonly agreed to be true. Some of its predictions have been successfully tested. We are confident that it accurately maps between conscious states and physical states on platinum standard systems during consciousness experiments. We can use this *reliable* c-theory to make inferences about the consciousness of decapitated heads, bats and robots.

In an experiment on a platinum standard system a c-theory's predictions can be checked because assumptions A1-A6 hold and we can measure consciousness. These assumptions do not apply to decapitated heads, bats and robots. We cannot obtain a believable c-report from these systems, so we cannot compare a c-description generated by a c-theory with a c-description generated from a c-report. Inferences about the consciousness of these systems cannot be confirmed or refuted. This type of untestable prediction will be referred to as a *deduction*, which is defined as follows:

> **D16**. A *deduction* is a c-description that is generated from a p-description when consciousness cannot be measured. Deductions are blind logical consequences of a c-theory. They cannot be tested because assumptions A1-A6 do not apply. The plausibility of a deduction is closely tied to the reliability of the c-theory that was used to make it.

Predictions are testable. Deductions are not. However much data I gather about a physical system I cannot *ever* test the deductions that I make about its consciousness. The assumptions that enable us to measure consciousness do not apply to the systems that we make deductions about.

I grab a bat, measure its physical state, and use a mathematical c-theory to convert a p-description of its physical state into a c-description of its consciousness (see Figure 9.3). If the bat's consciousness is radically different from my own, then it will be difficult for me to understand this c-description. I might find it impossible to imagine what it is like to be this bat. (How could I imaginatively transform my bubble of experience into the bat's bubble of experience?)[7] While solutions to this problem have been put forward,[8] at some point we will have to accept that we have a limited ability to imaginatively transform our bubbles of experience. This failure of imagination does not affect our ability to make scientific deductions about a bat's consciousness. A reliable c-theory should be able to generate a complete and accurate c-description of a bat's conscious state from a p-description of its physical state.[9]

There are strong ethical motivations for making deductions about the consciousness of brain-damaged people, embryos and infants. Deductions have implications for abortion, organ donation and the treatment of the dead and dying. We could use deductions to reduce the suffering of animals that are raised and slaughtered for meat.[10] Deductions could satisfy our curiosity about the consciousness of artificial systems.

Deductions will be based on c-theories that have not been fully tested. Poor access to the brain and limited time and money hamper our ability to test c-theories. We cannot check that a c-theory holds across all conscious and physical states. Multiple competing c-theories might be consistent with the evidence and exhibit different trade-offs between simplicity and generality. These problems are common to all scientific theories.

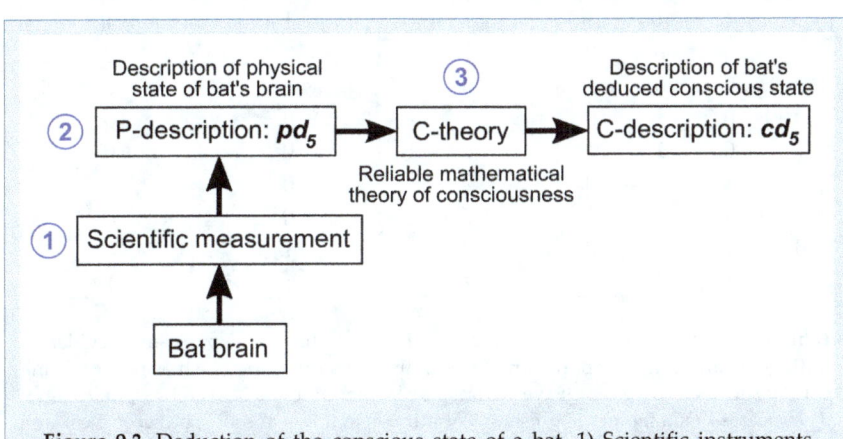

Figure 9.3. Deduction of the conscious state of a bat. 1) Scientific instruments measure the physical state of the bat's brain. 2) The measurements are converted into a formal p-description of the physical state of the bat's brain, pd_5. 3) A reliable well-tested c-theory converts the p-description into a formal c-description, cd_5, of the bat's deduced conscious state. Image © David Gamez, CC BY 4.0.

Deductions will be based on c-theories that are impossible to fully test. C-theories can only be tested in natural experiments on platinum standard systems. Under these conditions it will be difficult or impossible to prove that a c-theory is based on minimal sets of spatiotemporal structures. So there are likely to be residual ambiguities about CC sets that cannot be experimentally resolved. This is illustrated in Table 9.1.

Spatiotemporal structures				Results of experiments on platinum standard systems	Deductions about c_7 in a non-platinum standard system	
A	B	C	D	Conscious state c_7	Deductions of t_1 based on {B,C}	Deduction of t_2 based on {B,C,D}
0	0	0	0	?	0	0
0	0	0	1	0	0	0
0	0	1	0	?	0	0
0	0	1	1	0	0	0
0	1	0	0	?	0	0
0	1	0	1	0	0	0
0	1	1	0	?	1	0
0	1	1	1	1	1	1
1	0	0	0	?	0	0
1	0	0	1	0	0	0
1	0	1	0	?	0	0
1	0	1	1	0	0	0
1	1	0	0	?	0	0
1	1	0	1	0	0	0
1	1	1	0	?	1	0
1	1	1	1	1	1	1

Table 9.1. Deductions about consciousness based on limited experimental evidence. A, B and C are spatiotemporal structures in the physical world, such as neuron firing patterns or electromagnetic waves. D is a passive material, such as cerebrospinal fluid.[11] '1' indicates that a feature is present; '0' indicates that it is absent. The second column presents the results of experiments on platinum standard systems in which different combinations of A, B, C and D are tested and conscious state c_7 is measured. '1' indicates that c_7 is present; '0' indicates that c_7 is absent. The shaded rows are physical states in which D is absent. In this example D cannot be removed from a platinum standard system in a natural experiment, so the link between D and consciousness is unknown. A question mark in these rows indicates that it is not known whether c_7 is present when D is absent. On the basis of this data we can develop two c-theories, t_1 and t_2. t_1 links B and C to c_7; t_2 links B, C and D to c_7. Both of these theories are compatible with the experimental data. They make the same deductions about c_7 in the white rows and different deductions about c_7 in the shaded rows.

Many biological systems are similar to normally functioning adult human brains. With these systems we do not have to worry about whether a c-theory includes all of the spatiotemporal structures that *might* be linked to consciousness, because the normally functioning adult human brain and the target system contain similar patterns in similar materials. This will be expressed using the notion of a physical context:

> **D17.** A *physical context* is everything in a system that is not part of a CC set that is used to make a deduction. Two physical systems have the same physical context if they contain approximately the same materials and if the constant and partially correlated patterns in these materials are approximately the same.[12]

When two systems share the same physical context we can make deductions about their consciousness that are as strong and believable as the original theory. These will be referred to as *conservative* deductions, which are defined as follows:

> **D18.** In a *conservative* deduction a c-theory generates a c-description from a p-description in the *same* physical context as the one in which the theory was tested.

Suppose a c-theory links some electromagnetic patterns to consciousness. In this case the physical context is everything in the brain apart from these electromagnetic patterns, such as neurons, haemoglobin, cerebrospinal fluid, glia, other electromagnetic patterns, and so on. If these patterns and materials are approximately the same in another system, then they provide the same physical context for the electromagnetic patterns that the c-theory uses to make its deductions about consciousness.[13]

Other systems, such as cephalopods and robots, lack some of the spatiotemporal structures that are present in normally functioning adult human brains. The link between these spatiotemporal structures and consciousness cannot be tested in natural experiments. We can still make deductions about the consciousness of these systems, but they are likely to be less accurate than conservative deductions. These will be referred to as *liberal* deductions, which are defined as follows:

> **D19.** In a *liberal* deduction a c-theory generates a c-description from a p-description in a *different* physical context from the one in which the theory was tested.[14,15]

Suppose we have identified a mathematical relationship between neuron firing patterns and conscious states in the normally functioning adult human brain, and none of the brain's other materials need to be included to make accurate predictions about consciousness. We could use this mathematical relationship to make *conservative* deductions about the consciousness of a damaged brain, an infant's brain and possibly a bat's brain. In these brains, most of the same patterns and materials are present, so we would not have to prove that the c-theory is based on minimal sets of spatiotemporal structures. We could also use this relationship to make *liberal* deductions about the consciousness of neurons in a Petri dish or about a snail's consciousness. We would have less confidence in these deductions because these physical contexts lack some of the materials that are potential members of CC sets.

In the system described in Table 9.1, t_1 and t_2 make *conservative* deductions about consciousness in the physical states that are coloured white. These deductions are made in the same physical context as the one in which the theories were tested (D is always present), and both theories make identical deductions. t_1 and t_2 make *liberal* deductions about the physical states in the shaded rows. The theories have not been tested in this physical context, which lacks D, and they make different deductions. These liberal deductions cannot be confirmed or refuted—all that can be confirmed or refuted is the c-theory on which the deductions are based. For example, we might give preference to the liberal deductions of t_2 if it has performed better in experiments on platinum standard systems.

All well-tested c-theories should make the same conservative deductions. If two c-theories make different conservative deductions, then it should be possible to devise an experiment that can discriminate between them.

There can be contradictions between the liberal deductions that are made by equally reliable c-theories (see Table 9.1). Our reaction to this will depend on our motivation for making the deductions. If they are made out of interest, then we can say that the system is conscious according to t_1 and not conscious according to t_2. If they are made for ethical reasons, then we could base our treatment of the system on whether it has been deduced to be conscious according to *any* reliable c-theory. An artificial intelligence should not be switched off if it has been deduced to be conscious according to t_1, even if t_2 claims that it is unconscious.

The distinction between conservative and liberal deductions can be dropped if assumptions A7-A9 are made and a c-theory is considered to be true given these assumptions. In this case, the presence or absence of a physical context does not matter because passive materials, constant patterns and partially correlated patterns have been assumed to be irrelevant to consciousness. All deductions would then be equally valid.

9.3 Summary

A c-theory can generate testable predictions about consciousness or the physical world. Testable predictions can only be made about platinum standard systems during consciousness experiments. It is only under these conditions that we can use measurements of consciousness to generate or confirm the predictions.

When a c-theory has been rigorously tested, we might judge that it can reliably map between c-descriptions and p-descriptions. We can then use it to make deductions about the consciousness of non-platinum standard systems, such as coma patients and bats. These deductions cannot be checked because we cannot measure consciousness in these systems. We make deductions for a variety of ethical, practical and intellectual reasons.

Conservative deductions are made in the same physical context as the one in which the c-theory was tested. They are as reliable as the c-theory and all c-theories should make the same conservative deductions. Liberal deductions are made about systems that are substantially different from the platinum standard systems on which the c-theory was tested. They are less reliable than conservative deductions and different c-theories are likely to make different liberal deductions.

10. Modification and Enhancement of Consciousness

> […] our normal waking consciousness, rational consciousness as we call it, is but one special type of consciousness, whilst all about it, parted from it by the filmiest of screens, there lie potential forms of consciousness entirely different. We may go through life without suspecting their existence; but apply the requisite stimulus, and at a touch they are there in all their completeness…
>
> William James, *The Varieties of Religious Experience*[1]

10.1 Heaven on Earth

When we have understood the relationship between consciousness and the physical world we will be able to systematically modify and enhance our consciousness. We will achieve heaven on Earth without leaving home.

Why blow yourself up for Allah when you can deflower ten virgins per hour in a scientifically constructed consciousness? Or you could stuff your face with roast pig without feeling sated or sick. You could dress in cloth of gold and drink from diamond cups without rising from your silver bed. Or give free reign to your wrath and watch your schoolmaster being rogered with a red hot iron while badly-dressed dwarves bludgeon your boss to death. The pain of envy would be eliminated in a consciousness in which you are supreme dictator of the world—a consciousness rich with the sensation of a vast and satisfying pride.

The meek and mild might prefer the less earthy pleasures of prudence, justice, temperance, courage, faith, hope and charity. Or scientists could engineer mystical ecstatic experiences in which acolytes are penetrated by darts of divine love. Family audiences might enjoy the rich radiance of a sunset or the emotions induced by the birth of a child.

The modification and enhancement of consciousness has medical applications. Some people have damaged consciousness, a low level of consciousness or no consciousness at all. Other bubbles of experience are full of demons and intrusive thoughts. Some people have a sad sagging consciousness that they seek to escape through death. Many consciousnesses are permeated with relentless agonizing pain. A scientific approach to the modification and enhancement of consciousness would enable us to fix damaged consciousnesses, treat people with depression and eliminate pain. Some of the people who are diagnosed as schizophrenic might benefit from adjustments to their consciousness.

The modification and enhancement of consciousness could increase our empathy. I could experience things from your point of view. My consciousness could be merged with your consciousness when we make love.[2]

10.2 Types of Modification and Enhancement

Virtually every aspect of our bubbles of experience can be changed. Some of the main modifications are as follows:

- *Level of intensity.* The average level of intensity of a bubble of experience can be increased or decreased as well as the level of intensity of particular contents (Figure 10.1b).[3]
- *Contents.* The contents of bubbles of experience vary widely. There are bubbles of experience filled with black limitless space and bubbles of experience filled with dirty headless singing chickens (Figure 10.1c).[4]
- *Body location.* The location of our bodies in our bubbles of experience can be altered without changing the location of our physical bodies. This is known as an out-of-body experience. Suppose I am standing on a cliff looking out to sea. Without changing the location of my physical body, I can relocate my

body in my bubble of experience, so that I am floating in the air and looking back at myself on the cliff (Figure 10.1d).[5]

- *Body size.* The size of my body in my bubble of experience can be varied so that I am as small as a flea or as tall as the trees (Figure 10.1e).[6]
- *Body shape.* I can become a crow or grow an extra head (Figure 10.1f).[7]
- *Emotions.* The intensity of emotions can be increased or decreased.[8]
- *Space.* Our bubbles of experience can be expanded or contracted to hold more or fewer things in greater or less detail (Figure 10.1g).[9]
- *Time.* The present moment has a temporal thickness (the specious present), which could be expanded or contracted. Our short term memory could be increased or reduced and we could enhance our access to previous events (long term memory).[10]
- *Novel sensations.* Our bubbles of experience are limited to five or six senses. It might be possible to experience novel sensations.[11]
- *Mystical states.* Many of the states described by mystics can be interpreted as variations of the modifications that have already been described. For example, if our sense of body ownership is extended to our entire bubble of experience, then we experience a profound sense of oneness with our environment. A glowing vision of Jesus can be added to a bubble of experience. Mystical journeys can be interpreted as modifications of body location and contents. Our bubbles of experience could also be modifiable in completely novel and unimaginable ways (see Section 10.5).

None of these modifications and enhancements involve spooky stuff. They can all be brought about by changes to the physical brain.

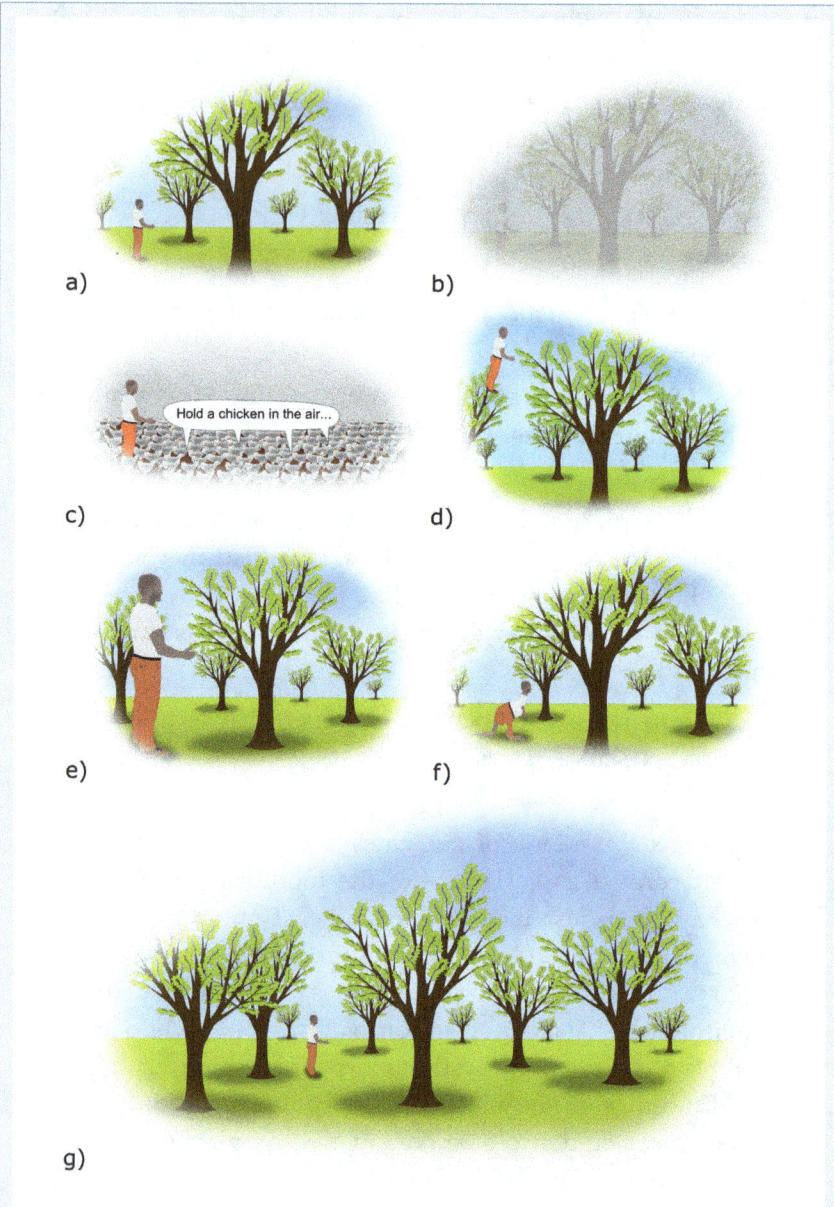

Figure 10.1. Modifications of a bubble of experience. a) Bubble of experience whose associated CC set is determined by sensory input from a lightly wooded landscape. b) Reduction of the average level of intensity. c) Change in contents. d) Change in the location of the body. e) Increase in the size of the body. f) Change in body shape. g) Spatial expansion of bubble of experience. Image © David Gamez, CC BY 4.0.

10.3 Scientific Modification and Enhancement of Consciousness

We modify our consciousness all the time by changing sensory input, imagining and ingesting chemicals. These techniques require no knowledge of the relationship between consciousness and the physical world.[12]

Scientific research on consciousness will enable us to modify and enhance our consciousness. Once we have identified the relationship between consciousness and the physical world, we can use this knowledge to create desired states of consciousness. This is a multi-stage process:

1. Generate a c-description, cd_6, of the desired state of consciousness.
2. Use a reliable c-theory to convert cd_6 into a p-description, pd_6, of the associated CC set.
3. Realize this CC set in the human brain.

This is illustrated in Figure 10.2.

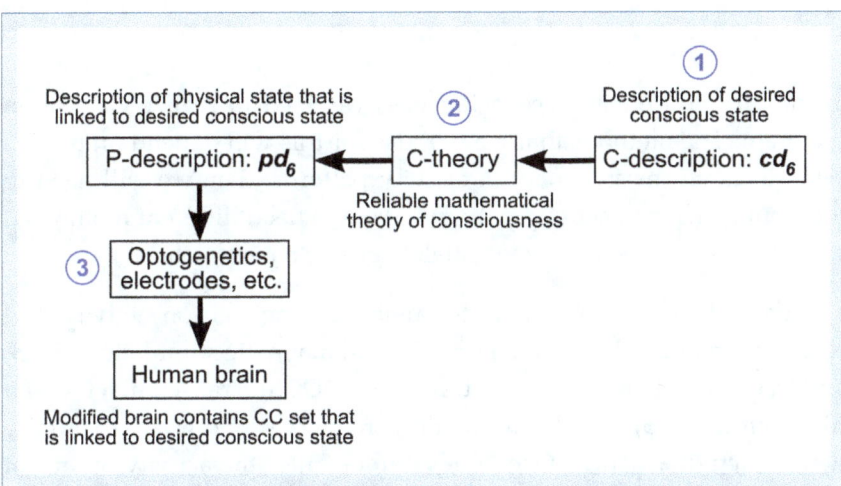

Figure 10.2. A reliable c-theory is used to realize a desired state of consciousness. 1) The desired state of consciousness is specified in a formal c-description, cd_6. 2) A reliable c-theory converts the c-description into a formal description of a physical state, pd_6. 3) The brain is modified using optogenetics, electrodes, etc. (see Section 10.4) so that it contains the CC set described in pd_6. Image © David Gamez, CC BY 4.0.

Suppose I want to sleep with the Queen. First I generate a formal detailed description of this state of consciousness. Then I use a reliable c-theory to convert this c-description into a p-description of the state of the physical world (the CC set) that is associated with this state of consciousness. Finally I use the methods described in Section 10.4 to put my brain into this state. I realize my dream—I am tucked up in bed in my PJs with the Queen.[13]

This approach could be used to modify an animal's consciousness. My guinea pig's conscious body could be enhanced with an extra leg. Or I could reduce its pain when I exploit it for meat.

This technology could be commercialized. In the distant future we might have designers of consciousness, who work with a customer to generate a c-description of the consciousness they want to achieve. The designers would then realize the corresponding CC set in the customer's brain. People could experience the consciousness of Jenna Jameson or John Malkovich. Instead of watching a film, we could experience it from a first-person perspective—we would really feel the actors' pains and pleasures.[14]

10.4 Methods

To modify and enhance consciousness we need to realize CC sets in the brain. The methods that we will use for this will depend on the CC sets—if they consist of neuron activity patterns, then we will need to manipulate neuron activity patterns. We will need different methods if CC sets contain electromagnetic fields, glia or haemoglobin.[15]

The non-invasive methods for manipulating neuron activity and electromagnetic fields include transcranial magnetic stimulation (TMS) and transcranial direct current stimulation (tDCS). Transcranial focused ultrasound (tFUS) uses the mechanical effects of sound waves to modify neuron activity.[16] These techniques crudely alter the activity of tens of thousands of neurons, so they are unlikely to play much of a role in the modification of consciousness based on c-theories.

Invasive technologies provide detailed control over the firing behaviour of individual neurons. Electrodes can control up to a hundred neurons at a time;[17] optogenetics can potentially control thousands of neurons.[18] In the longer term nanotechnology might lead to higher resolution methods for brain control.[19]

Chemicals are usually delivered to the brain through the blood, which exposes the entire brain to the chemical. In the future it might be possible to target chemicals more precisely. We could develop drugs that are specific to CC sets, inject chemicals directly into the brain or genetically engineer neurons to make them more selectively responsive to chemicals.

More tissue could be added to the brain,[20] which could self-organize in response to stimulation patterns. Synthetic neurons could be implanted (if they were valid members of CC sets).[21] These could have enhanced properties, such as a higher firing rate.[22]

Some of these methods might require implanted silicon chips.[23] These would not form part of the CC sets or be associated with consciousness by themselves. The link between implanted electronics and consciousness is part of the research on machine consciousness, which is covered in the next chapter.

We are a long way from realizing specific CC sets in the human brain. It is possible that the technology for realizing CC sets will have substantially improved by the time that we have reliable c-theories and good formats for c-description and p-description.[24]

Some modifications of the human brain can be done in natural experiments. For example, many foods and most sensory inputs do not jeopardize the status of normally functioning adult human brains as platinum standards. If A1-A6 apply to the modified brain, then we can measure its consciousness to check that we have created the desired bubble of experience.

Other methods preserve the physical context of the brain that is being modified. For example, optogenetics and electrodes modify the activity of a small number of neurons and have little effect on the rest of the brain. If the physical context is preserved, we will be able to *conservatively deduce* that the desired state of consciousness is present.

Some methods change the physical context when they realize a CC set in the brain. These include chemicals delivered through the blood and crude methods for modifying brain activity, such as TMS, tDCS and tFUS. Under these conditions we can only make *liberal deductions* about the presence of a desired state of consciousness.

10.5 Beyond What We Can Imagine

> It is difficult, it is all but impossible, to speak of mental events except in similes drawn from the more familiar universe of material things. If I have made use of geographical and zoological metaphors, it is not wantonly, out of a mere addiction to picturesque language. It is because such metaphors express very forcibly the essential otherness of the mind's far continents, the complete autonomy and self-sufficiency of their inhabitants. A man consists of what I may call an Old World of personal consciousness and, beyond a dividing sea, a series of New Worlds—the not too distant Virginias and Carolinas of the personal subconscious and the vegetative soul; the Far West of the collective unconscious, with its flora of symbols, its tribes of aboriginal archetypes; and, across another, vaster ocean, at the antipodes of everyday consciousness, the world of Visionary Experience.
>
> Aldous Huxley, *Heaven and Hell*[25]

The Romans could have built steam engines, but they had no idea about this technology. They did not imagine it and did not build it. When I was two I had no inkling about my adult life. We cannot imagine the consciousness of fish or bats.

We use previous experiences to imagine how our consciousness could change. But the most interesting modifications and enhancements probably cannot be imagined by us. Mystics and hippies have peered into these realms. Many more states and modifications might be possible.

C-descriptions can help us to understand what lies beyond the limits of our imagination. If we had a good c-description format, we would be able to generate c-descriptions of all possible states of consciousness. We might be able to glimpse aspects of them in virtual reality. To enter these unknown regions we need reliable c-theories and better methods for realizing CC sets in the brain.

10.6 Summary

When we have reliable c-theories we will be able to modify and enhance our consciousness in different ways. Eventually we will be able to write down a c-description of a desired state of consciousness, use a reliable c-theory to map the c-description onto a p-description, and then modify the human brain so that the subject experiences the desired state of consciousness.

At the present time we do not have reliable c-theories and we have not solved the problems of c-description and p-description. We have a very limited ability to realize CC sets in the brain. The scientific modification and enhancement of consciousness has great potential, but we might have to wait 50, 500 or 500,000 years.

11. Machine Consciousness

To actually create a technical model of full blown, perspectivally organized conscious experience seems to be the ultimate technological utopian dream. It would transpose the evolution of mind onto an entirely new level [...]. It would be a historical phase transition. [...] But is this at all possible? It certainly is conceivable. But can it happen, given the natural laws governing this universe and the technical resources at hand?

Thomas Metzinger, *Being No One*[1]

"Could a machine think?" My own view is that only a machine could think, and indeed only very special kinds of machines, namely brains and machines that had the same causal powers as brains. And that is the main reason strong AI has had little to tell us about thinking, since it has nothing to tell us about machines. By its own definition, it is about programs, and programs are not machines. [...] No one would suppose that we could produce milk and sugar by running a computer simulation of the formal sequences in lactation and photosynthesis, but where the mind is concerned many people are willing to believe in such a miracle because of a deep and abiding dualism: the mind they suppose is a matter of formal processes and is independent of quite specific material causes in the way that milk and sugar are not.

John Searle, Minds, Brains and Programs[2]

11.1 Types of Machine Consciousness

A team of scientists labours to build a conscious machine. They ignite its consciousness with electricity and it opens its baleful eye. It declares that it is conscious and complains about its inhuman treatment. The scientists liberally deduce that it is *really* conscious. They run tests to probe its reactions to fearful stimuli. Terrified, it snaps its chains and runs amok in the lab. It rips one intern apart and bashes out the brains of another. With a wild rush it bursts through the door and disappears into the night.

© David Gamez, CC BY 4.0   http://dx.doi.org/10.11647/OBP.0107.11

This machine exhibited conscious external behaviour and really was conscious. It could have been controlled by a model of a CC set or a model of phenomenal consciousness. These different types of machine consciousness will be labelled MC1-MC4:[3]

- *MC1. Machines with the same external behaviour as conscious systems.* Humans behave in particular ways when they are conscious. They are alert, they can respond to novel situations, they can inwardly execute sequences of problem-solving steps, they can execute delayed reactions to stimuli, they can learn and they can respond to verbal commands (see Section 4.1). Many artificially intelligent systems exhibit conscious human behaviours (playing games, driving, reasoning, etc.). Some people want to build machines that have the full spectrum of human behaviour.[4] Conscious human behaviours can be exhibited by systems that are not associated with bubbles of experience.[5]

- *MC2. Models of CC sets.* Computer models have been built of potential CC sets in the brain.[6] This type of model can run on a computer without a bubble of experience being present. A model of a river is not wet; a model of a CC set would only be associated with consciousness if it produced appropriate patterns in appropriate materials. This is very unlikely to happen when a CC set is simulated on a digital computer.

- *MC3. Models of consciousness.* Computer models of bubbles of experience can be built.[7] These could be based on the phenomenological observations of Husserl, Heidegger or Merleau-Ponty. These models can be created without producing patterns in materials that are associated with bubbles of experience.

- *MC4. Machines associated with bubbles of experience.* When we have understood the relationship between consciousness and the physical world we will be able to build artificial systems that are actually conscious. These machines would be associated with bubbles of experience in the same way that human brains are associated with bubbles of experience. Some of our current machines might already be associated with bubbles of experience.

Several different types of machine consciousness can be present at the same time. We can build machines with the external behaviour associated with consciousness (MC1) by modelling CC sets or consciousness (MC2, MC3).[8] We could produce a machine that exhibited conscious external behaviour (MC1) using a model of CC sets (MC3) that was associated with a bubble of experience (MC4).

The construction of MC1, MC2 and MC3 machines is part of standard computer science. The construction of MC4 machines goes beyond computer models of external behaviour, CC sets and consciousness. MC4 machines contain patterns in materials that are associated with bubbles of experience.

11.2 How to Build a MC4 Machine

MC4 machine consciousness would be easy if computations or information patterns could form CC sets by themselves. Unfortunately computations and information patterns do not conform to constraints C1-C4, so they cannot be used to build MC4 machines (see Chapters 7 and 8).

To construct a MC4 machine we need to realize particular patterns in particular physical materials. If we had a reliable c-theory, we could design and build a MC4 machine in the following way:

1. Generate c-description, cd_7, of the consciousness that we want in the machine.
2. Use reliable c-theory to convert cd_7 into a p-description, pd_7, of the CC set that corresponds to this conscious state.
3. Realize this CC set in a machine.

This is illustrated in Figure 11.1.

We do not have a reliable c-theory. So we can only guess about the patterns and materials that might be needed to build MC4 machines.

If CC sets contain patterns in electromagnetic fields, then we could use neuromorphic chips to generate appropriate electromagnetic patterns in an artificial system.[9] If CC sets contain biological neurons, then we could build an artificial MC4 system using cultured biological neurons.[10]

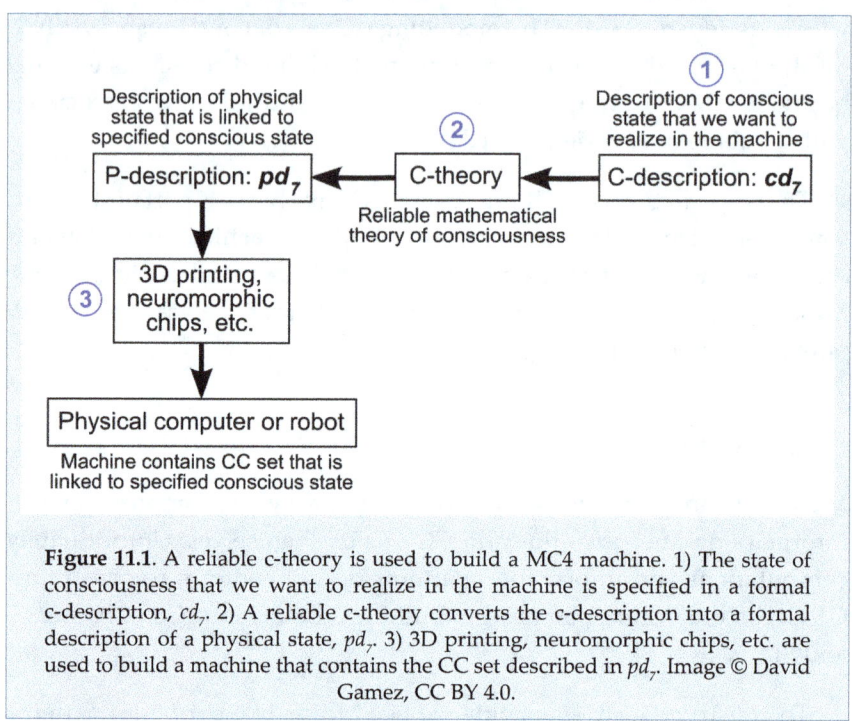

Figure 11.1. A reliable c-theory is used to build a MC4 machine. 1) The state of consciousness that we want to realize in the machine is specified in a formal c-description, cd_7. 2) A reliable c-theory converts the c-description into a formal description of a physical state, pd_7. 3) 3D printing, neuromorphic chips, etc. are used to build a machine that contains the CC set described in pd_7. Image © David Gamez, CC BY 4.0.

11.3 Deductions about the Consciousness of Artificial Systems

Last year I purchased a C144523 super-intelligent mega-robot from our local store. It cooks, cleans, makes love to the wife and plays dice. Last week I saw a new model in the shop window—it is time to dispose of C144523. On the way to the dump it goes on and on about how it is a really sensitive robot with real feelings. It looks sad when I throw it in the skip. The wife is sour. Little Johnny screams 'How can you do this to C144523? She was conscious, just like us. I hate you, I hate you, I *hate* you!' Perhaps C144523 really had real feelings? I probably should have checked.

We want to know whether the MC1-MC3 machines that we have created are really conscious. We want to know whether we have built a MC4 machine.

We can use reliable c-theories to make deductions about the MC4 consciousness of artificial systems. These are likely to be liberal deductions because most machines do not have the same physical context as our current platinum standard systems (see Section 9.2).

Suppose we have a reliable c-theory that maps electromagnetic patterns onto conscious states. We would analyze an artificial system for MC4 consciousness in the following way:

1. Measure its electromagnetic patterns.
2. Convert measurement into p-description, pd_g.
3. Use reliable c-theory to convert pd_g into a c-description, cd_g, of the artificial system's consciousness.

This is illustrated in Figure 11.2. The plausibility of the resulting c-description depends on the reliability of the physical c-theory and on whether it is a conservative or a liberal deduction.[11]

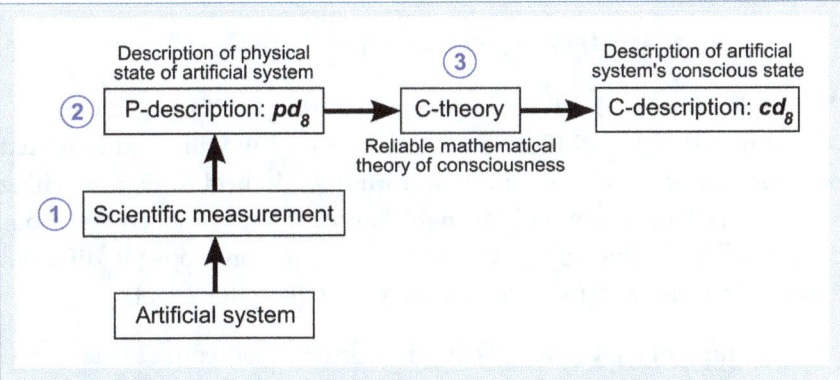

Figure 11.2. A reliable c-theory is used to deduce the consciousness of an artificial system. 1) Scientific instruments are used to measure the physical state of the artificial system. 2) Scientific measurements are converted into a formal p-description of the artificial system's physical state, pd_g. 3) A reliable c-theory converts the p-description into a formal description of the artificial system's conscious state, cd_g. Image © David Gamez, CC BY 4.0.

11.4 Limitation of this Approach to MC4 Consciousness

This approach to MC4 machine consciousness is based on c-theories that are developed using platinum standard systems. But our platinum standard systems might not contain all of the patterns and materials that are linked to bubbles of experience.

One type of pattern and material could be linked to consciousness in the human brain; a different type of pattern and material could be linked to consciousness in an artificial system. It is impossible to find out whether this is the case. We cannot measure the consciousness of systems that are not platinum standards, so we cannot *prove* that the patterns in their materials are *not* associated with conscious states. At best we can be confident that a machine *is* MC4 conscious—we cannot be confident that it is *not* MC4 conscious.

In the future we might be seduced by a machine's MC1 behaviour and assume that it is a platinum standard system. However, assumptions about platinum standard systems should be not made lightly—they have radical implications for the science of consciousness (see Section 5.4).

11.5 Conscious Brain Implants

Artificial devices could be implanted in our brains to extend our consciousness.[12] In a MC4 implant the CC set would be distributed between the brain and the implant, forming a hybrid human-machine system. MC4 implants would enable us to modify and enhance our consciousness more easily. We could become conscious of different types of information (from our environment, the Internet, etc.).

MC4 implants have medical applications. CC sets could be damaged by tumours, strokes or accidents. The damaged area could be replaced with an implant that was deduced to have the missing consciousness.

MC4 implants would have to produce specific patterns in specific materials. The patterns could be distributed between the brain and the implant. For example, if CC sets consisted of electromagnetic patterns, then neuromorphic chips[13] could be implanted, which would work together with the brain's biological neurons to create electromagnetic patterns that would be associated with consciousness.

Brains with implants are not platinum standard systems, so we cannot measure their consciousness using c-reports. Only conservative or liberal deductions can be made about their consciousness.

11.6 Uploading Consciousness into a Computer

It will soon be technologically possible to scan a dead person's brain and create a simulation of it on a computer.[14] Some people think that a simulation will have the same consciousness as their biological brain. They believe that they can achieve immortality by uploading their consciousness into a computer.[15]

It is extremely unlikely that a simulation of your brain on a digital computer will have a bubble of experience. Simulations have completely different electromagnetic fields from real brains and lack the biological materials that might be members of CC sets.

Your consciousness can only be uploaded into an artificial system that reproduces the CC sets in your brain. We do not know which of the brain's materials are present in CC sets, so the only certain way of uploading your consciousness is to create an atom-for-atom copy of your brain.

The advantage of the brain-uploading approach is that the patterns linked to consciousness are blindly copied from the original brain. Suppose we could show that electromagnetic fields are the only materials in CC sets. I could then upload my consciousness by realizing my brain's electromagnetic field patterns in a machine. This could be done without any knowledge of the patterns that are linked to consciousness.

When I upload a file to the Internet the file remains on my computer. The same would be true if I uploaded my consciousness by scanning my living brain using non-destructive technology. *My* consciousness would continue to be associated with *my* biological brain. A *copy* of my consciousness would be created in the computer. Scanning and simulating my brain would not *transfer* my consciousness—it would not enable *my* consciousness to survive the death of my biological body.[16]

11.7 Will Conscious Robots Conquer the World?

Science fiction fans know that conscious machines will take over the world and enslave or eliminate humans. Prescient science fiction writing helped us to prepare for the Martian invasion. Perhaps we should take drastic action *now* to save humanity from conscious killer robots?[17]

The next section argues that a takeover by superior MC1 or MC4 machines could be a good thing. This section examines the more common (and entertaining) belief that humanity will be threatened by malevolent machine intelligences.

MC1 machines carry out actions in the world—they fire lasers, hit infants on the head and steal ice cream. Research on MC2, MC3 and MC4 systems might improve our ability to develop MC1 machines, but MC2, MC3 and MC4 machines cannot achieve anything unless they are capable of external behaviour. Only MC1 machines could threaten humanity.[18]

It is extremely difficult to develop machines with human-level intelligence. We are just starting to learn how to build specialized systems that can perform a single task, such as driving or playing Jeopardy. This is much easier than building MC1 machines that learn as they interact with the world and exhibit human-like behaviour in complex dynamic environments.[19]

Let us take the worst-case scenario. Suppose super-intelligent computers control our aircraft, submarines, tanks and nuclear weapons. There are billions of armed robots. Every aspect of power generation, mining and manufacturing is done by robots. Humans sit around all day painting and writing poetry. Under these conditions MC1 machines could take over. However, if humans were involved in the manufacture and maintenance of robots, if they managed the mines and power production, then complete takeover is very unlikely—the robot rebellion would rapidly grind to a halt as the robots ran out of power and their parts failed.

People who worry about machines taking over should specify the conditions under which this would be possible.[20] When we get close to fulfilling these conditions we should take a careful look at our artificially intelligent systems and do what is necessary to minimize the threat.

Some people have suggested that artificial intelligence could run away with itself—we might build a machine that constructs a more intelligent machine that constructs a more intelligent machine, and so on. This is known as a *technological singularity*.[21] The fear that we might build a machine that takes over the world is replaced by a higher-order fear that we build a machine that builds a machine that builds a machine that takes over the world. We have little idea how to build such a machine.[22] Just a sickening sense of *fear* when we *imagine* an evil super-intelligence hatching from a simple system.

Suppose we write an intelligent program that writes and executes a more intelligent program, and so on at an accelerating rate. What can this super-intelligent system *do*? In what way could it pose an existential threat to humanity? Physically it can do nothing unless it is connected to a robot body. And what can one robot do against ten billion humans? On the Internet the super-intelligence will be able to do everything that humans do (make money through gambling, hire humans to do nefarious deeds, purchase weapons, etc.). But it is very unlikely to pose more of a threat than malevolent humans. Large, well-funded teams of highly intelligent humans struggle to steal small sums of money, copy business secrets, and carry out physical and online attacks. There is little reason to believe that a super-intelligent system could achieve much more. A runaway intelligence would only pose a threat if many other conditions were met. We are very far from building this type of system and we will have plenty of time to minimize the risks if it becomes a real possibility.[23]

Machines are much more likely to *accidentally* destroy humanity as a result of hardware or software errors. Killer robots with appropriate kill switches might make better decisions on the battlefield and cause less collateral damage. However, if we have thousands of such robots, then there is a danger that a software error could kill the kill switch and set them on the rampage. Protecting humans against software errors is not straightforward because most modern weapons systems are under some form of computer control and humans can make calamitous decisions based on incorrect information provided by computers.[24]

Science fiction reflects our present concerns—it tells us little about a future that is likely to happen. Over the next centuries our attitudes towards ourselves and our machines will change. As our artificial intelligences improve we will get better at understanding, regulating and controlling them. Po-faced discussions about the existential threat of artificial intelligence will become as quaint as earlier fears about Martian invaders.

11.8 Should Conscious Robots Conquer the World?

> It is very unlikely that intelligent machines could possibly produce more dreadful behaviour towards humans than humans already produce towards each other, all round the world even in the supposedly most civilised and advanced countries, both at individual levels and at social or national levels.
>
> Aaron Sloman, Why Asimov's Three Laws of Robotics Are Unethical[25]

Many humans are stupid useless dangerous trash. Scum rises to the top. People have killed hundreds of millions of people in grubby quests for power, pride, sexual satisfaction and cash. We have come close to nuclear catastrophe.

Super-intelligent MC1 machines might run the world better than ourselves and make humanity happier. They could systematically analyze more data without the human limitations of boredom and self-interest. MC1 machines could maximize human wellbeing without petty political gestures.[26]

Positive states of consciousness have value in themselves. We fear death because we fear the permanent loss of our consciousness. Crimes are ethically wrong because of their effects on conscious human beings.[27] Nothing would matter if we were all zombies.

Many human consciousnesses are small and mediocre. People pour out unclean and lascivious thoughts to their confessors. We are guilty of weak sad thoughts and pathetically wallow in negative states of consciousness. Human bubbles of experience fall far short of consciousness' potential.

Reliable c-theories would enable us to engineer MC4 machines with better consciousness. If consciousness is ethically valuable in itself, then a takeover by superior MC4 machines would be a good thing. This could be a gradual process without premature loss of life that we would barely notice. We would end up with 100 billion high quality machine consciousnesses, instead of 10 billion human consciousnesses full of hate, lies, gluttony and war. A pure brave new world without sin in thought or deed.[28]

We have species-specific prejudices, a *narrow-mindedness*, that causes us to recoil with shock and horror from the suggestion that we should meekly step aside in favour of superior machines. This does not mean that the ethical argument is *wrong* or that the replacement of humans with MC4 machines would be *bad*—merely that it is against the self interest of our species. Looked at dispassionately a good case can be made. Can we rise above our prejudices and create a better world?

The world might be a better place if MC4 machines took over. But it is very unlikely that this utopian scenario will come to pass. The science of consciousness has a long way to go before we will be able to design better consciousnesses. And we are only likely to be able to make *liberal* deductions about the consciousness of artificial systems, which are not completely reliable. We should only replace humans with MC4 machines when we are certain that the machines have superior consciousnesses.

11.9 The Ethical Treatment of Conscious Machines

> What would you say if someone came along and said, "Hey, we want to genetically engineer mentally retarded human infants! For reasons of scientific progress we need infants with certain cognitive and emotional deficits in order to study their postnatal psychological development— we urgently need some funding for this important and innovative kind of research!" You would certainly think this was not only an absurd and appalling but also a dangerous idea. It would hopefully not pass any ethics committee in the democratic world. However, what today's ethics committees *don't* see is how the first machines satisfying a minimally sufficient set of constraints for conscious experience could be just *like* such mentally retarded infants. They would suffer from all kinds of functional and representational deficits too. But they would now also subjectively experience those deficits. In addition, they would have no political lobby—no representatives in *any* ethics committee.
>
> Thomas Metzinger, *Being No One*[29]

The science of consciousness should enable us to build MC4 machines that are associated with bubbles of experience. Some of our MC1, MC2 or MC3 machines could already be MC4 conscious. These systems might suffer; they might be confused; they might be incapable of expressing their pain.

We want machines that exhibit the behaviours associated with consciousness (MC1). We want to build models of CC sets (MC2) and models of consciousness (MC3). But we might have to *prevent* our machines from becoming MC4 conscious if we want to avoid the controversy associated with animal experiments.

It would be absurd to give rights to MC1, MC2 or MC3 machines. Ethical treatment should be limited to machines that are really MC4 conscious.

We can use reliable c-theories to deduce which machines are MC4 conscious (see Section 11.3). One potential problem is that a large number of physical objects (phones, toasters, cars etc.) might be deduced to be MC4 conscious according to the most reliable c-theory. It is also highly unlikely that we will reach the stage of designing systems with zero or positive states of consciousness without building systems that have 'retarded' or painful consciousness.

11.10 Summary

There are four types of machine consciousness. There are machines whose external behaviour is similar to conscious systems (MC1), there are models of CC sets (MC2), models of consciousness (MC3), and systems that are associated with bubbles of experience (MC4). Reliable c-theories can be used to deduce which machines are really MC4 conscious. Some implants and brain scanning/upload methods are potentially forms of MC4 machine consciousness.

The production of conscious machines raises ethical questions, such as the potential danger to humanity of MC1 machines, whether MC1 or MC4 machines should take over the world, and how we should treat MC4 machines.

It could take hundreds or thousands of years to develop artificial systems with human levels of consciousness and intelligence. It might be impossible to build super-intelligent machines. Current discussion of these issues is little more than speculation about a distant future that we cannot accurately imagine.

12. Conclusion

Science does not rest on solid bedrock. The bold structure of its theories rises, as it were, above a swamp. It is like a building erected on piles. The piles are driven down from above into the swamp, but not down into any natural or 'given' base; and if we stop driving the piles deeper, it is not because we have reached firm ground. We simply stop when we are satisfied that the piles are firm enough to carry the structure, at least for the time being.

Karl Popper, *The Logic of Scientific Discovery*[1]

12.1 A Framework for the Science of Consciousness

This book has set out a systematic framework for the scientific study of consciousness. It has tried to shift consciousness research into a paradigmatic state.[2] The key points are as follows:

- *Consciousness is a bubble of experience*. It consists of colours, sounds, smells, tastes, etc., which are arranged in a bubble of space centred on our bodies.
- *The physical world is invisible*. It has none of the secondary qualities that are present in a bubble of experience. Primary qualities in a bubble of experience are unlikely to resemble primary qualities in the physical world.
- *There are three hard problems of consciousness*. First, it is impossible to imagine the relationship between consciousness and the invisible physical world. Second, we find it difficult to imagine the connection between conscious experiences of brain activity and other conscious experiences. Third, there are brute regularities between consciousness and the physical

world that cannot be broken down or further explained. None of these issues affect scientific research on consciousness.

- To *scientifically study consciousness* we need to measure consciousness, measure the physical world and look for mathematical relationships between these measurements.
- Consciousness is measured through the external behaviour (*c-reports*) of systems that we assume to be conscious (*platinum standard systems*).
- Normally functioning adult human brains are platinum standard systems.
- Measurements of consciousness need to be expressed in a formal way (a *c-description*), so that they can be incorporated into mathematical theories of consciousness.
- The correlates of a conscious state are a set of spatiotemporal structures in the physical world (a *CC set*) that is only present when the conscious state is present. A CC set is *functionally connected to a bubble of experience* and it *e-causes c-reports about the bubble of experience*.
- CC sets need to be described in a formal context-free way (a *p-description*), so that they can be incorporated into mathematical theories of consciousness.
- *Pilot studies* attempt to identify the CC sets that are associated with individual conscious states.
- *C-theories* are mathematical relationships between c-descriptions of conscious states and p-descriptions of CC sets.
- *Computational methods* should be used to discover c-theories.
- *Information c-theories* claim that information patterns could form CC sets by themselves. *Computation c-theories* claim that computations could form CC sets by themselves. These types of c-theory do not conform to the constraints on scientific theories of consciousness (C1-C4).

- *Physical c-theories* link patterns in particular materials to conscious states. They conform to the constraints on CC sets and fit in well with other scientific theories.
- C-theories can generate *predictions* about consciousness or the physical world. Predictions can only be tested on platinum standard systems.
- C-theories can make *conservative and liberal deductions* about the consciousness of non-platinum standard systems, such as bats, infants and robots. Deductions are logical consequences of a c-theory that cannot be checked.

This framework handles or sets aside most of the philosophical problems with consciousness.[3] We cannot solve these problems. We can only show that they are pseudo problems, suspend judgement about them or set them aside with assumptions.

This framework is compatible with some of the metaphysical approaches to consciousness, such as physicalism and epiphenomenalism. It suspends judgement about which (if any) are correct. It is not compatible with panpsychism, dualism,[4] or with information and computation c-theories.

This framework prescribes the *form* that legitimate theories of consciousness should take. Our final c-theories will not be lengthy pieces of natural language. They will be mathematical relationships between c-descriptions and p-descriptions.

This framework is neutral about which physical c-theory is correct. While I have often used neurons and electromagnetic fields as examples, I have no idea about which patterns and materials are actually linked to consciousness. This question should be addressed by scientific experiments.

A person who accepts this framework can focus on measuring consciousness, measuring the physical world and identifying the relationships between these measurements. Their results can be considered to be true given assumptions A1-A6—they cannot be obtained or justified without these assumptions.[5]

12.2 Technological Limitations

The science of consciousness can only fully develop when we have accurate high resolution measurements of consciousness and the physical world.

The problems with measuring consciousness are mostly conceptual and methodological (see Chapter 4). If enough effort is spent, we should be able to obtain detailed and reasonably complete measurements of conscious states.

To accurately measure the physical world we need p-description methods that avoid the heavy reliance on context that is common in biology (see Section 5.1). We also need better access to the 100 billion neurons in the living human brain. The most commonly used technologies are as follows:

- *Functional Magnetic Resonance Imaging (fMRI)*. An indirect measure of brain activity with a spatial resolution of a few thousand voxels. Each voxel corresponds to the activity of approximately 50,000 neurons averaged over several seconds.
- *Diffusion Magnetic Resonance Imaging (dMRI)*. Identifies structural connections between brain areas, but does not show their direction. Can only help to identify CC sets when combined with other methods.
- *Electroencephalography (EEG)*. Approximately 300 electrodes are placed on the scalp to measure the brain's electrical field with good temporal resolution and very poor spatial resolution.
- *Magnetoencephalography (MEG)*. Measures the magnetic fields generated by groups of 50,000 neurons at approximately 300 points on the head with good temporal resolution.
- *Implanted electrodes*. Up to 300 electrodes can be implanted in the brain to measure electromagnetic fields and neuron activity with good spatial and temporal resolution.[6] Electrodes are rarely implanted in human subjects.
- *Optogenetics*. Neurons can be genetically engineered to emit light when they fire, which enables their individual activity to

be recorded using light sheet microscopy. This technique can be used to record from 100,000 neurons in a zebrafish larva in close to real time.[7] It is more challenging to use optogenetics in mammalian brains and for ethical reasons it has not been used on human subjects.

The data that is extracted using these techniques can be processed into higher level properties. For example, we can use Granger causality or dynamic causal modelling to identify effective connections between brain areas. These connectivity patterns can be further analyzed using graph theory.[8]

Optogenetics is the most promising technology for obtaining high resolution data from living brains. However, there are ethical and safety concerns about using it on humans. To get around these problems we can use animal brains to make inferences about CC sets in humans. Or we can assume that monkeys and mice are platinum standard systems.[9]

C-theories can be based on patterns that have higher resolution than our current measurement technologies. For example, we could develop c-theories based on neuron activity patterns and predict how these neuron activity patterns would appear in EEG data or a fMRI scan.

The scanning and uploading of a human brain could help to address our measurement problems. We could identify the structures in a simulated brain that cause its simulated c-reports. This might help us to develop c-theories that are not limited by our current technologies.

12.3 Other Limitations

I did not get my picture of the world by satisfying myself of its correctness; nor do I have it because I am satisfied of its correctness. No: it is the inherited background against which I distinguish between true and false.

Ludwig Wittgenstein, *On Certainty*[10]

The framework presented in this book cannot be shown to be correct. It is a condition of possibility of experimental work on consciousness that cannot be verified by experimental work on consciousness.[11] Scientific research within this framework might be fruitful and yield reliable c-theories. Or a science of consciousness based on this framework might reach a point at which it no longer coherently hangs together. We might have to formulate a completely new set of framing principles. Or we might have to abandon the attempt to scientifically study consciousness.

The inappropriate use of intuition, thought experiments and imagination has led to many problems in the philosophy of consciousness. I have tried to banish these as much as possible, but they cannot be completely eliminated. For example, I have assumed that normally functioning adult human brains are platinum standard systems. But which systems count as normally functioning adult human brains? What counts as a legitimate chemical modification? There are no natural boundaries—we have to use our intuition and imagination to decide which human brains are platinum standard systems.

This framework leaves many questions unanswered. It does not explain what consciousness *is*, what consciousness *does*, what consciousness is *for*, how consciousness *arose*, *why* there is a functional connection between consciousness and the physical world or *how* this connection actually works.

These questions are about consciousness *in general*. But it makes no sense to ask about the physical world *in general*. We can ask about the origin and function of particular physical structures—we cannot meaningfully ask about the origin and function of the entire universe. The case is similar with consciousness—it is meaningless to ask most of these questions about consciousness in general.

Let's rephrase these questions. Consider a state of consciousness, c_g. c_g is a bubble of experience in which you are peeping through a hole at an old woman in a bath. We can ask what c_g is, what c_g does, what c_g is *for*, how c_g arose, *why* c_g is functionally connected to the physical world and *how* this connection actually works.

These questions can be answered if we assume that c_g is identical to its associated CC set, cc_g. This explains the connection between c_g and the physical world. As the science of consciousness progresses we will get a better understanding of what cc_g is, what cc_g does and how it arose through e-causal processes, such as evolution. All of our questions about c_g can be answered by transferring them to cc_g.

Our questions about c_g can also be answered by assuming that it is a teapot. This tells us what c_g is (a teapot), how it arose (a factory in China) and what it is for (making tea). But assumptions about consciousness have to be *valid*, they have to *make sense*. An assumption's validity has to be decided independently of our desire to obtain cheap and easy answers about consciousness. No answers are better than bad answers.

It makes little sense to say that colourful smelly noisy bubbles of experience are *identical* to something that is invisible, silent and without smell. This discards the properties of bubbles of experience and ignores the reality of our day-to-day world. We could equally well discard the physical world and declare that it is a fairytale told by simple folk to explain regularities in consciousness.[12] Neither reduction is part of the framework that is set out in this book. Consciousness and the physical world are both taken as basic realities that can be measured and scientifically studied.

The physical sciences' assumption that the physical universe exists leaves many questions unanswered. Many questions will remain unanswered if conscious states are not reduced to physical states. We will simply have to accept that consciousness exists and study the brute regularities between consciousness and the physical world.

12.4 Future Research

If we take in our hand any volume; of divinity or school metaphysics, for instance; let us ask, Does it contain any abstract reasoning concerning quantity or number? No. Does it contain any experimental reasoning concerning matter of fact and existence? No. Commit it then to the flames: For it can contain nothing but sophistry and illusion.
    David Hume, *An Enquiry Concerning Human Understanding*[13]

The following types of consciousness research are likely to be productive:
- *Revision of the assumptions*. We might be able to reduce the number of assumptions, improve their consistency and

enhance the way in which they relate to general principles in the philosophy of science and the study of consciousness.

- *Creation of c-description format.* We need a precise formal way of describing states of consciousness that can be incorporated into mathematical c-theories.
- *Improvement of methods for measuring consciousness.* More work is required on how we can obtain detailed measurements of conscious states.
- *Creation of a context-free p-description format.* We need a formal context-free way of describing biological structures, such as neurons.
- *Development of a precise definition of a physical context.* Conservative and liberal deductions can only be distinguished when we have a precise definition of a physical context.
- *Increase the spatial and temporal resolution of our brain measurements.* We can refine existing methods, develop new technologies and create better mathematical techniques for processing data into higher level properties.
- *Pilot studies on the correlates of consciousness.* More pilot studies could help us to identify the patterns and materials that form CC sets.
- *Development of physical c-theories.* In the medium to longer term we need to move beyond pilot studies and identify mathematical relationships between bubbles of experience and physical states.[14]
- *Experimental testing of physical c-theories.* Physical c-theories will only be considered to be reliable when their predictions have been experimentally confirmed.
- *Construction of computer models of CC sets that simulate c-reports.* This could help us to identify the patterns and materials that form CC sets. These models could also be used to develop methods for the computational discovery of c-theories.
- *Development of methods for the computational discovery of c-theories.* This could apply existing work on the computational

12. Conclusion

discovery of scientific theories to data from consciousness and the brain.

- *Deductions about the consciousness of non-platinum standard systems.* When we have a reliable c-theory we will be able to answer questions about the consciousness of bats, infants and robots. Deductions also have important medical applications.
- *Experimental work on the modification and enhancement of consciousness.* When we have reliable c-theories and better technology for modifying the brain we will be able to systematically modify and enhance human consciousness.
- *Construction of MC1-MC4 machines.* This has many practical applications and could be a useful way of studying human consciousness.

A book or paper on consciousness that describes none of these things should be committed to the flames. Or carefully checked for sophistry and illusion.

Appendix: Definitions, Assumptions, Lemmas and Constraints

Definitions

D1. *Consciousness* is another name for *bubbles of experience*. A state of a consciousness is a state of a bubble of experience. Consciousness includes all of the properties that were removed from the physical world as scientists developed our modern invisible explanations (2.5).[1]

D2. A *c-report* is a physical behaviour that is interpreted as a report about a person's consciousness (4.1).

D3. A *nc-report* is a physical behaviour that is interpreted as a report about non-conscious mental content (4.2).

D4. A *platinum standard system* is a physical system that is assumed to be associated with consciousness some or all of the time (4.3).

D5. A *correlate of conscious state* is a minimal set of one or more spatiotemporal structures in the physical world. This set is present when the conscious state is present and absent when the conscious state is absent. This will be referred to as a *CC set* (4.6).

D6. A *c-description* is a formal description of a conscious state (4.9).

D7. A *p-description* is a formal description of a spatiotemporal structure in the physical world. A p-description unambiguously determines

---

1   The number in brackets is the section in which the definition, assumption, lemma or constraint can be found.

whether a spatiotemporal structure is present in a sequence of physical states (5.1).

D8. In a *natural experiment* the test system preserves its status as a platinum standard. Assumptions A1-A6 remain valid and consciousness can be measured throughout the experiment (5.4).

D9. In an *unnatural experiment* the test system is transformed into something that is not a platinum standard. A1-A6 cease to apply and we lose our ability to measure the system's consciousness (5.4).

D10. A *c-theory* is a compact expression of the relationship between consciousness and the physical world. A c-theory can generate a c-description from a p-description, and generate a p-description from a c-description (5.5).

D11. A *material* is an arrangement of elementary wave-particles at a particular spatial scale (6.1).

D12. A *physical c-theory* links consciousness to spatiotemporal patterns in materials. Physical CC sets consist of one or more patterns and the materials in which these patterns occur (6.1).

D13. An *information c-theory* links consciousness to spatiotemporal information patterns. Information CC sets only contain information patterns, which can occur in any material (7.2).

D14. A *computation c-theory* links consciousness to the execution of computations. Computation CC sets only contain computations, which can be executed by many different types of computer (8.2).

D15. A testable *prediction* is a c-description that is generated from a p-description or a p-description that is generated from a c-description. Predictions can be checked by measuring consciousness or they are generated from measurements of consciousness. Predictions can only be generated or confirmed on platinum standard systems during experiments on consciousness. It is only under these conditions that consciousness can be measured using assumptions A1-A6 (9.1).

D16. A *deduction* is a c-description that is generated from a p-description when consciousness cannot be measured. Deductions are blind logical consequences of a c-theory. They cannot be tested because assumptions A1-A6 do not apply. The plausibility of a deduction is closely tied to the reliability of the c-theory that was used to make it (9.2).

D17. A *physical context* is everything in a system that is not part of a CC set that is used to make a deduction. Two physical systems have the same physical context if they contain approximately the same materials and if the constant and partially correlated patterns in these materials are approximately the same (9.2).

D18. In a *conservative* deduction a c-theory generates a c-description from a p-description in the *same* physical context as the one in which the theory was tested (9.2).

D19. In a *liberal* deduction a c-theory generates a c-description from a p-description in a *different* physical context from the one in which the theory was tested (9.2).

Assumptions

A1. During an experiment on consciousness, the consciousness associated with a platinum standard system is functionally connected to the platinum standard system's c-reports (4.3).

A2. During an experiment on consciousness *all* conscious states associated with a platinum standard system are available for c-report and all aspects of these states can potentially be c-reported (4.3).

A3. The consciousness associated with a platinum standard system nomologically supervenes on the platinum standard system. In our current universe, physically identical platinum standard systems are associated with indistinguishable conscious states (4.4).

A3a. The bubble of experience that is associated with a CC set nomologically supervenes on the CC set. In our current universe, physically identical CC sets are associated with indistinguishable conscious states (4.6).

A4. The normally functioning adult human brain is a platinum standard system (4.5).

A5. The physical world is e-causally closed (4.7).

A6. CC sets e-cause a platinum standard system's c-reports (4.7).

A6a. CC sets are effectively connected to a platinum standard system's c-reports (4.7).

A7. *CC sets do not contain passive materials.* If the link between consciousness and the simple presence of a material cannot be demonstrated in a natural experiment, then this material can be excluded from potential CC sets (6.4).

A8. *CC sets do not contain patterns that are present when the system is conscious and unconscious.* If the link between consciousness and a constant pattern cannot be demonstrated in a natural experiment, then this pattern can be excluded from potential CC sets (6.4).

A9. *CC sets do not contain partially correlated patterns.* When several different materials have the same spatiotemporal pattern, the material(s) in which the spatiotemporal pattern is strongest will be considered to be the potential member(s) of the CC set that is associated with the conscious state, unless the partially correlated patterns can be separated out in a natural experiment (6.4).

Lemmas

L1. There is a functional connection between a conscious state and its corresponding CC set (4.6).

Constraints

C1. *The spatiotemporal structures in a CC set are independent of the observer.* My consciousness is a real phenomenon that does not depend on someone else's subjective interpretation. CC sets must be formed from objective spatiotemporal structures, such as electromagnetic waves and neuron firing patterns (5.2).

C2. *The members of CC sets are intrinsic properties.* A conscious state supervenes on a CC set (A3a), so each duplicate of a CC set must be associated with an identical conscious state, regardless of the spatial and temporal context in which the duplicate appears (5.2).

C3. *A non-conscious system does not contain a CC set that is 100% correlated with a conscious state*. If A and B are 100% correlated, then A cannot occur without B. If a CC set is 100% correlated with a conscious state, then all brains that contain that CC set will be conscious (5.2).

C4. *CC sets e-cause c-reports during consciousness experiments (A6)*. It is not necessary for every member of a CC set to e-cause c-reports. But some parts or aspects of the CC set must e-cause them. So when I say 'I am conscious of a green tomato', this c-report can be traced back to the CC set that e-caused it, which is functionally connected to a bubble of experience in which there is a green tomato (5.2).

Endnotes

2. The Emergence of the Concept of Consciousness

1. Qualia (singular: quale) is a technical philosophical term that refers to the qualitative or subjective properties of experiences. For exaple, the colour red, the taste of chocolate and the sound of a bell are qualia.

2. See Gamez (2007, Chapter 2) and Lehar (2003) for more detailed descriptions of bubbles of perception. Husserl (1964) has a good analysis of the temporal structures of our bubbles of perception.

3. It might be thought that a bubble of experience is a version of Dennett's (1992) Cartesian Theatre, in which a homunculus observes the contents of consciousness. This appears to require a second homunculus inside the head of the first, and so on ad infinitum. However, bubbles of experience do not have the same problems as Cartesian Theatres because there is no perception within a bubble of experience. My physical brain perceives its physical environment using electromagnetic waves, etc.; my conscious experience of my body does not perceive my conscious experience of my environment using a conscious experience of electromagnetic waves — there is no transmission of information within a bubble of experience. I have discussed this in more detail elsewhere (Gamez 2007, pp. 47-8).

4. For example, Lucretius (2007).

5. Locke (1997, p. 137).

6. This example is taken from ancient scepticism. Studies have shown that people have different perceptions of bitterness (Hayes et al. 2011) and there is substantial variability in people's olfactory perception (Mainland et al. 2014).

7. See Galilei (1957) and Locke (1997). This distinction was also developed by the ancient atomists — see Taylor (1999) for a discussion.

8. It might be claimed that honey is sweet in itself and produces sensations of sweetness (or bitterness) through the interaction of its sweetness with

our senses. In this case the different properties perceived by different observers would be due to the different ways in which the physical sweetness interacts with their senses. The problem with this proposal is that it is impossible to decide whether the honey is sweet and produces a false bitter sensation when it interacts with Zampano's senses, or whether it is bitter and produces a false sweet sensation when it interacts with my senses. It is much simpler to attribute all sensory properties to the interaction between the physical world and the sense organs.

9   Locke (1997, pp. 136-7).

10   For example, O'Regan and Noë claim: 'There can therefore be no one-to-one correspondence between visual experience and neural activations. Seeing is not constituted by activation of neural representations. Exactly the same neural state can underlie different experiences, just as the same body position can be part of different dances.' (O'Regan and Noë 2001, p. 966). A less radical position can be found in Noë's later work: 'A reasonable bet, at this point, is that some experience, or some features of some experiences, are, as it were, exclusively neural in their causal basis, but that full-blown, mature human experience is not.' (Noë 2004, p. 218).

11   This is formally stated as assumption A4 in Section 4.5.

12   Noë's (2004) bet that this type of conscious experience is not exclusively correlated with neural activity is a different working assumption that can be experimentally tested. I have discussed this point in more detail elsewhere (Gamez 2014b).

13   When a neuron fires it emits a short electrical pulse known as an action potential or spike. This electrical pulse has an amplitude of ~100 mV, a duration of ~2ms and it can be transmitted to other neurons or passed along the nerves.

14   As Dennett puts it: 'The representation of space in the brain does not always use space-in-the-brain to represent space, and the representation of time in the brain does not always use time-in-the-brain.' (Dennett 1992, p. 131). The distinction between space and time in the mind and space and time in the objective world was introduced by Kant (1996), who claimed that space and time are forms of intuition. According to Kant it is unknowable whether space and time are present in the objective noumenal world.

15   It might be thought that the traditional primary property of number is an exception. However, number is not a physical property, but the magnitude of a physical property, which is obtained through a measurement procedure and varies with the system of units. For example, we can carry

out an act of counting that results in a number, or extract the ratio of two masses as a number. Consider a ball that weighs 7.3 kilos (16.1 pounds): the mass is a physical property of the ball, not the numbers 1, 7.3 or 16.1.

16 Kant's (1996) metaphysics expresses a similar idea: the noumenal physical world is an invisible source of signals that are processed through the categories to become phenomenal experiences.

17 This use of discrete black boxes to illustrate objects in the physical world is not strictly correct because the boundaries between physical objects depend on the observer's sensory apparatus and ontology (Gamez 2007, Chapter 5).

18 Russell (1927, p. 163).

19 Many people today have a different interpretation of our bubbles of experience, which is often aligned with idealism and rejects the scientific interpretation of physical reality—Tibetan Buddhism is one example. This book will not examine these other interpretations of consciousness and the physical world.

20 'Consciousness' is sometimes used to refer to an individual person's consciousness and sometimes used as a mass term to refer to all of the consciousness in existence—just like 'water' is used to refer to all of the water in existence. 'What is consciousness?' and 'What is water?' treat consciousness and water as mass terms. In this definition I have tried to limit the ambiguity by linking a state of *a* consciousness to a state of *a* bubble of experience.

21 Galilei (1957, p. 274).

22 Suppose a person's bubble of experience contains three objects, P, Q and R. P appears with intensity 0.7, Q appears with intensity 0.8 and R appears with intensity 0.3 (these values are purely illustrative). According to the proposed definition, this person's overall level of consciousness would be the average of these intensity values: (0.7+0.8+0.3)/3 = 0.6.

23 This is similar to our use of 'awake', except a person can be awake without having a bubble of experience. For example, vegetative state patients are presumed to be unconscious, but they can have cycles of wakefulness in which they open their eyes and move their body in meaningless ways (Laureys et al. 2002).

24 Metzinger (2003) has a good discussion of online and offline conscious experience.

25 Functional connectivity (a deviation from statistical independence between A and B) is typically contrasted with structural connectivity (a physical link between A and B) and from effective connectivity (a causal link from A to B)—see Friston (1994; 2011). A number of algorithms exist for measuring functional and effective connectivity.

26 The research on change blindness, attentional blindness and change detection in peripheral vision suggests that the amount of online conscious content is less than we think (Cohen and Dennett 2011; Rensink et al. 1997; Simons and Chabris 1999; Simons and Rensink 2005).

27 Wilkes (1988b, pp. 16-7).

28 Wilkes (1988b, p. 38).

29 Lucretius (2007) claims that the soul (a combination of spirit [*anima*, the vital principle] and mind [*animus*, the intellect]) is a subtle particle. See Chapter 6, Footnote 10.

3. The Philosophy and Science of Consciousness

1 See Husserl (1960).

2 For example, Smart (1959).

3 See Section 2.1.

4 For example, piano-playing pigs are unlikely to have entered the Aztecs' imaginations.

5 See Nagel (1974).

6 Other problems with thought experiments and imagination have been discussed by Wilkes (1988a) and Gamez (2009). The *Stanford Encyclopedia of Philosophy* has a good overview (Brown and Fehige 2014).

7 McGinn (1989, p. 349).

8 Metzinger (2000, p. 1).

9 The quote by McGinn at the beginning of this section is a typical description of the hard problem of consciousness. Chalmers (1995b) made a popular distinction between easy and hard problems of consciousness. Strawson (2015) gives a historical overview.

10 This example has been simplified. If the brain-imaging device showed the complete state of my brain on the screen, then the conscious experience

of p_1 would be linked to everything that was going on in my brain, both consciously and unconsciously—it would not just be the pattern associated with my conscious experience of the ice cube. Multiple experiments would be required to identify and selectively display the brain activity that was linked to my conscious experience of the ice cube.

11 To make the text easier to read I have separated conscious experiences of brains from other conscious experiences. However, our conscious experience of a brain is a conscious experience. So in this example, some of the brain patterns on the screen will be associated with our conscious experience of the brain patterns on the screen.

12 Our finite cognitive capacities (long and short term memory, etc.) will limit our ability to learn associations between conscious experiences of brain patterns and other conscious experiences. For example, it is unlikely that we will be able to learn all of the details of a complex brain pattern.

13 Rorty (1980, p. 71).

14 I have told this story from the perspective of consciousness. It can also be told in terms of physical brain activity. If we knew enough about the brain, we could describe how it learns to associate sensory stimuli from brains with other sensory stimuli.

15 Elementary wave-particles and superstrings are only put forward as examples. Future advances in physics might explain the behaviour of elementary wave-particles and superstrings in terms of brute regularities at a lower level.

16 Boyle's law is a good example of a scientific law that can be explained in terms of regularities at a lower level. It states that the pressure of a gas is inversely proportional to its volume in a closed system at constant temperature. This macro-scale experimental observation can be explained in terms of the behaviour of atoms and molecules, which was formerly treated as a brute regularity that was the starting point for scientific explanations.

4. The Measurement of Consciousness

1 Chalmers (1998, p. 220).

2 Descriptions of consciousness can be interpreted as statements about the physical world. When I report that I have a conscious experience of a rusty helmet beside my conscious experience of my left foot, I am also reporting that there is a rusty helmet beside my left foot in the physical world.

3   Wittgenstein (1969) discusses how our knowledge is underpinned by a framework of certainties that cannot be doubted without putting everything into question.

4   When we imagine different motor tasks, such as walking around a house or playing tennis, we activate different brain areas that can be discriminated in a fMRI scanner. This enables people to answer yes/no questions about their consciousness by imagining that they are performing one of two actions. This method has been used to communicate with patients in vegetative or minimally conscious states, who were incapable of other forms of voluntary behaviour (Monti et al. 2010; Owen et al. 2006).

5   This list of behaviours includes suggestions from Shanahan (2010), Koch (2004) and Teasdale and Jennett (1974).

6   Post-decision wagering is a method that is used to measure consciousness in psychology (Persaud et al. 2007). A person is asked to make a decision and to bet on the accuracy of that decision. It is assumed that the person will bet more money on decisions that are based on conscious information. See Sandeberg et al. (2010) for a comparison of post-decision wagering, the perceptual awareness scale and confidence ratings.

7   An overview of some of the techniques for measuring consciousness is given by Seth et al. (2008).

8   Damasio (1999, p. 6).

9   An overview of binocular rivalry is given by Blake (2001).

10 This is a simplified summary of the large number of experiments that have been carried out on visual masking and non-conscious perception. For example, Dell'Acqua and Grainger (1999) showed that unconsciously perceived pictures influenced subjects' ability to consciously name pictures and categorize words. Schütz et al. (2007) showed that masked prime words can influence how subjects complete gap words. Merikle and Daneman (1996) played words to patients under general anaesthesia and found that when they were awake they completed word stems with words that they had heard non-consciously. A change in the skin's conductivity is known as a galvanic skin response, which can indicate that information is being processed unconsciously (Kotze and Moller 1990). Öhman and Soares (1994) showed that subjects' skin conductance response changed when they unconsciously perceived phobic stimuli, such as pictures of snakes or spiders. A review of experimental work on visual masking and non-conscious perception is given by Kouider and Dehaene (2007).

11 This is known as forced choice guessing. While some people believe that above chance results on a forced choice guessing task demonstrate that conscious information is present, blindsight patients can guess the identity of visual stimuli above chance while reporting no subjective awareness (Weiskrantz 1986). Seth et al. (2008) discuss these issues.

12 By 'associated' it is meant that consciousness is linked to a platinum standard system, but no claims are being made about causation or metaphysical identity.

13 The metre used to be defined as one ten-millionth of the distance from the Earth's equator to the North Pole at sea level. Since this was difficult to measure, a platinum-iridium bar was used instead. Rulers were directly or indirectly calibrated against this bar, which was kept in Paris.

14 If the platinum-iridium standard metre doubled in size, an object that used to be 1 metre long (1 platinum-iridium standard metre bar) would have a new length of 0.5 metres (0.5 platinum-iridium standard metre bars). This would only be strictly true if the platinum-iridium bar was the actual definition of the metre, rather than the working definition. The same argument applies to the actual definition of the metre.

15 Functional connectivity (a deviation from statistical independence between A and B) is typically contrasted with structural connectivity (a physical link between A and B) and from effective connectivity (a causal link from A to B)—see Friston (1994; 2011). A number of algorithms exist for measuring functional connectivity (for example, mutual information), and it can be measured with a delay.

16 While phenomenal consciousness and access 'consciousness' *might* be conceptually dissociable (Block 1995), the idea that non-measureable phenomenal consciousness could be present during experiments on consciousness is incompatible with the scientific study of consciousness. Block's non-accessible phenomenal consciousness does not appear in c-reports, so everything that Block has ever written or said about it is meaningless or false.

17 A possible exception to this would be a situation in which non-reportable consciousness is present but does not interfere with our ability to identify the correlates of consciousness. This is discussed in more detail in Chapter 9, Footnote 13.

18 This is similar to Block's (2007) idea of cognitive accessibility.

19 Dennett questions the idea that there is a single stream of consciousness with a fixed content: 'the Multiple Drafts model avoids the tempting mistake of supposing that there must be a single narrative (the 'final or 'published' draft, you might say) that is canonical—that is the *actual* stream of consciousness of the subject, whether or not the experimenter (or even the subject) can gain access to it.' (Dennett 1992, p. 113). Personally I do not find Dennett's arguments for his multiple drafts model convincing. We will find out if he is correct, because it will be impossible to obtain systematic stable measurements of consciousness.

20 Panpsychism is the view that all matter is linked to consciousness. For example, some versions of panypsychism claim that individual electrons, quarks, etc. are associated with simple bubbles of experience.

21 A2 is also likely to be incompatible with Zeki and Bartels' (1999) proposal that micro-consciousnesses are distributed throughout the brain.

22 The perceived colour of an object does not just depend on the frequencies of the electromagnetic waves that are reflected, transmitted or emitted by it. Our visual system also uses the spectrum of illuminating light and the colour of surrounding objects to identify an object's colour, which enables us to attribute the same colour to objects under different lighting conditions. I have used electromagnetic wave frequencies to simplify the presentation of the colour inversion argument, which also applies to a more accurate account of colour perception.

23 There are likely to be subtle behavioural differences between two colour-inverted people—see Palmer (1999) for a discussion of behaviourally equivalent inversion scenarios. These differences would disappear if completely different sets of 'colour' experiences were linked to frequencies of electromagnetic waves.

24 If a set of properties, A, supervenes on another set of properties, B, then it is impossible for two things to have different A properties without also having different B properties. This is not a causal relationship.

25 Kouider et al. (2013) and Dehaene (2014) discuss infant consciousness.

26 Animal consciousness is discussed by Dehaene (2014), Edelman and Seth (2009) and Feinberg and Mallatt (2013).

27 People with Anton-Babinski syndrome are blind, but claim that they can see and confabulate to cover up the contradictory evidence. Other anosognosia patients are completely paralyzed on one side, but claim that their body is working perfectly.

28 Section 2.4 discusses theories that link consciousness to sensorimotor interactions between the brain, body and environment. In previous work I made the assumption that the *awake* normal adult human brain is a platinum standard system (Gamez 2011; Gamez 2012a). The more developed account of c-reports presented in this book makes the assumption that the brain is awake unnecessary—immobility, unresponsiveness, etc. are c-reports of zero consciousness.

29 Although brain-damaged patients have played an important role in consciousness research, they should not be uncritically assumed to be platinum standard systems. There can be ambiguities about whether the damage has knocked out the memory and reporting functions and left the consciousness intact, or knocked out the consciousness and left the memory and reporting functions intact. For example, locked-in patients are thought to be fully conscious, but they are only capable of moving their eyes, and some of the patients studied by Owen et al. (2006) and Monti et al. (2010) are likely to be conscious but unable to display this in their external behaviour. The use of brain-damaged patients in consciousness research has the further problem that the damage is typically non-localized and some brain areas are likely to perform several different functions. One way of addressing this issue is to assume that brain-damaged patients are platinum standard systems on a case-by-case basis, taking the details of the damage into account and its likely impact on memory and/or reporting.

A similar ambiguity applies to the use of anaesthetics in consciousness research. For example, midazolam, xenon and propofol are used to induce unconsciousness, so that scientists can compare the state of the conscious and unconscious brain (Casali et al. 2013). This raises the question whether the anaesthetic completely removes consciousness, or just paralyzes the body and prevents the subject from remembering and reporting their consciousness. This issue can also be addressed on a case-by-case basis. We can examine the mechanism of each anaesthetic and decide whether it is likely to affect the areas linked to memory and/or reporting. Normally functioning adult human brains containing anaesthetics that do not affect memory and/or reporting can be assumed to be platinum standard systems.

Animal experiments can also be handled on a case-by-case basis. We can assume that the brains of monkeys or mice are associated with consciousness, so that we can use these animals in consciousness research.

30 The notion of a minimal set is intended to exclude features of the brain that typically occur at the same time as consciousness, whose removal would not lead to the alteration or loss of consciousness. For example, a CC set

31 might have prerequisites and consequences (Aru et al. 2012; de Graaf et al. 2012) that typically co-occur with consciousness, but the brain would be conscious in exactly the same way if the CC set could be induced without these prerequisites and consequences.

31 This is similar to Chalmers' (2000) definition of the total correlates of consciousness, which he distinguishes from the core neural basis: 'A total NCC builds in everything and thus automatically suffices for the corresponding conscious states. A core NCC, on the other hand, contains only the "core" processes that correlate with consciousness. The rest of the total NCC will be relegated to some sort of background conditions required for the correct functioning of the core.' (Chalmers 2000, p. 26). Block (2007) makes a similar distinction.

32 This will not be correct if some spatiotemporal structures can *inhibit* consciousness. For example, we might have a CC set, cc_1, that is a correlate of consciousness according to D5. In most circumstances consciousness would be present whenever cc_1 was present. However, if consciousness was inhibited by ih_1, then there could be a situation in which cc_1 and ih_1 were present together and there was no consciousness.

33 Footnote 15 explains the relationship between functional and effective connectivity. These are typically inferred from data using algorithms, such as Granger causality or mutual information, and they are distinct from physical causation, which is discussed in the next section.

34 There are many spurious correlations—for example, see Vigen (2016). These can be divided into false correlations, which are the result of poor statistical procedures, and true but unlikely correlations that might be due to an underlying cause. When there is a true correlation between A and B it is possible to obtain information about B by measuring A and vice versa (the amount of information that one can obtain depends on the strength of the correlation). In this book I am presenting a framework that is based the assumption that there is a true statistical correlation (functional connection) between consciousness and the physical world. So we can obtain information about consciousness by measuring parts of the physical world and obtain information about the physical world by measuring consciousness.

35 Kim (1998, p. 31).

36 This distinction is taken from Dowe (2000). It is similar to Fell et al.'s (2004) distinction between efficient and explanatory causation. Efficient causation is concerned with the physical relation of two events and the exchange of physically conserved quantities. Explanatory causation refers to the law-like character of conjoined events.

37 Predominantly conceptual accounts of causation include Lewis' (1973) counterfactual analysis and Mackie's (1993) INUS conditions. Empirical theories based on the exchange of physically conserved quantities have been put forward by Aronson (1971a; 1971b), Fair (1979) and Dowe (2000). Bigelow et al. (1988) and Bigelow and Pargetter (1990) link causation to physical forces.

38 See Dowe (2000).

39 A world line is the path of an object through space and time.

40 If all empirical theories of causation are unworkable, then we might have to limit causal concepts to ordinary language and abandon the attempt to develop a scientific understanding of the causal relationship between consciousness and the physical world.

41 Kim (1998) has a good discussion of the relationship between macro and micro physical laws.

42 Wilson (1999) discusses the minimum amount of physical effect that would be required for consciousness to influence the physical brain.

43 A related point is made by Fell et al. (2004), who argue that the neural correlates of consciousness cannot e-cause conscious states.

44 Controversial experiments by Libet (1985) have indicated that our awareness of our decision to act comes after the motor preparations for the act (the readiness potential). This suggests that our conscious will might not be the cause of our actions, and Wegner (2002) has argued that we make inferences after the fact about whether we caused a particular action. These results could be interpreted to show that CC sets do not e-cause c-reports about consciousness because motor preparations for verbal output (for example) would precede the events that are correlated with consciousness. This problem could be resolved by measuring the relative timing of a proposed correlate of consciousness (CC1 in Figure 4.4) and the sequence of events leading to the report about consciousness, including the readiness potential (R1-R3 in Figure 4.4). If the framework presented in this book is correct, then it should be possible to find CC sets with the appropriate timing relationship. If no suitable CC sets can be found, then the framework presented in this book should be rejected as flawed. It is worth noting that Libet's measurement of the timing of conscious events implicitly depends on a functional connection between consciousness and c-reporting behaviour—the relative timing of consciousness and action can only be measured if consciousness is functionally connected to c-reports about consciousness (in this case with a delay).

45 It is reasonably easy to see how the contents of consciousness that are c-reported could be e-caused by physical events. For example, we can tell a simple story about how light of a particular frequency could lead to the activation of spatiotemporal structures in the brain, and how learning processes could associate these with sounds, such as 'red' or 'rojo'. This might eventually enable a trained brain to produce the sounds 'I can see a red hat' or 'I am aware of a red hat' when it is presented with a pattern of electromagnetic waves. Since consciousness does not appear to us as a particular thing or property in our environment and many languages do not contain the word 'consciousness' (Wilkes 1988b), it is not necessary to identify sensory stimuli that the physical brain could learn to associate with the sound 'consciousness'. The concept of consciousness can be more plausibly interpreted as an abstract concept that is acquired by subjects in different ways (see Chapter 2). So it is conceivable that the scientific study of consciousness could be carried out without subjects ever using the word 'consciousness' in their c-reports.

46 See Footnote 15 for the distinction between structural, functional and effective connectivity. Effective connectivity can be measured using algorithms, such as transfer entropy (Schreiber 2000) or Granger causality (Granger 1969), which works on the assumption that a cause precedes and increases the predictability of the effect. However, effective connectivity does not always coincide with e-causation—for example, when a signal is connected to two areas with different delays.

47 In the real brain many areas are reciprocally connected to each other and there is a great deal of recurrent processing. This simplified diagram only shows the general flow of activity from perception to reporting.

48 Cohen and Dennett (2011) illustrate the low resolution of our peripheral vision.

49 Our ability to access high resolution information on demand contributes to our sense that we perceive the world in uniformly high resolution (O'Regan 1992).

50 This is a conservative estimate based on eye-movement driven changes and the assumption that consciousness consists of a series of discrete moments (the specious present). It is also possible that consciousness changes continuously.

51 See O'Regan (1992).

52 People can be trained to make more accurate reports about their consciousness (Lutz et al. 2002) and there has been a substantial amount

of work on the use of interviews to help people describe their conscious states. In the explication interview (EI) a trained person interviews a subject about a conscious state to help them provide an accurate report (Petitmengin 2006). In descriptive event sampling (DES) the subject carries a beeper, which goes off at random several times per day. When they hear the beep the subject makes notes about their consciousness just before the beep. This is followed by an interview that is designed to help the subject to provide faithful descriptions of the sampled experiences (Hurlburt and Akhter 2006). Froese et al. (2011) discuss some of the first- and second-person methods for measuring consciousness. These techniques place a heavy reliance on memory, so it is unlikely that they can address the problems highlighted in this section.

53   Shanahan (2010) suggests how an omnipotent psychologist could measure a person's consciousness by reversing time and carrying out different interventions.

54   Mental techniques could also be used to reset consciousness. For example, people with a high level of mental focus, possibly gained through meditation, might be capable of putting their consciousness into a particular state and maintaining this state for an extended period of time.

55   It might be possible to use what we know about the relationship between a stimulus and consciousness to make reliable inferences about a person's state of consciousness. Suppose we knew that an awake expectant person always has a conscious experience of a red rectangle when a red rectangle is presented at the centre of their visual field. If this inference was reliable, it might not be necessary to measure their consciousness using c-reports when we expose them to a red rectangle—we could simply infer that since they are looking at a red rectangle, they must be conscious of a red rectangle. However, the limited resolution and active nature of the visual system means that a complex model will be required to map between stimuli and conscious states. Furthermore, this method of inference can only be developed by measuring consciousness using c-reports, which depends on the assumptions that have been presented in this chapter.

56   Nagel (1974, p. 449).

57   You might think that you could validate the descriptions by resetting your consciousness to the state that is being described. But then you would have to compare a remembered description with your current state of consciousness without modifying your current state of consciousness.

58   Formal descriptions of the physical world are covered in Section 5.1.

59 These problems are discussed by Chrisley (1995a) and Gamez (2006).

60 The use of XML to describe consciousness is discussed by Gamez (2006; 2008a).

61 See Balduzzi and Tononi (2009).

5. Correlates and Theories of Consciousness

1 This is the current definition of a metre.

2 This is our conscious experience of measurement. I could also describe how Randy's height is measured by the physical brain of the scientist.

3 Eddington (1928, pp. 251-2).

4 This definition of measurement is a simplified version of the one put forward by Pedhazur and Schmelkin (1991), who take it from Stevens (1968). According to Stevens, most measurement involves 'the assignment of numbers to aspects of objects or events according to one or another rule or convention.' (Stevens 1968, p. 850). Pedhazur and Schmelkin stress that numbers are assigned to *aspects* of objects, not to objects themselves. We measure the height, width and colour of a box, not the box *itself*.

5 For example, if we cannot devise a way of p-describing neurons, then it will be difficult to make inferences about the consciousness of animals with larger neurons, such as snails and insects, and we will not be able to say anything about the consciousness of artificial systems.

6 Intrinsic properties are tied to an object's physical nature. They are held by an object independently of its spatial and temporal context. Extrinsic properties depend on an object's relationships with other parts of the world. The chemical composition of a neuron is an intrinsic property. The distance of a neuron from the North Pole is an extrinsic property, which would change if the North Pole changed location.

7 It is conceivable that some CC sets could be 60% correlated with conscious states. Experimental work could determine whether this is the case. C3 will not apply if there are inhibitors of consciousness (see Chapter 4, Footnote 32).

8 Elsewhere I distinguished between type A and type B correlates of consciousness (Gamez 2014c). Type A correlates can e-cause c-reports and are compatible with C4. Type B correlates are not compatible with C4 because they cannot e-cause c-reports.

9   This discussion assumes that there is a 1:1 ratio between CC sets and conscious states.

10 The technologies that are available for measuring the brain are covered in Section 12.2.

11 I have discussed elsewhere how the correlates of consciousness can be experimentally separated out from their prerequisites and consequences and from sensory and reporting structures (Gamez 2014c). Pitts et al. (2014) describe experimental work that attempts to separate the correlates of conscious perception from reporting structures. This is also discussed by Koch et al. (2016).

12 Rees et al. (2002), Tononi and Koch (2008), Dehaene (2014) and Koch et al. (2016) describe some of the research that has been carried out on the neural correlates of consciousness.

13 Any kind of 'passive' monitoring or measurement involves the passage of physically conserved quantities from the system to the measuring device. In a natural experiment this is small compared to the exchange of physically conserved quantities within the system, so it does not affect our assumption that the system is a platinum standard.

14 This experiment has been extensively discussed—for example, by Moor (1988), Chalmers (1996a), Van Heuveln et al. (1998) and Prinz (2003). Part of the brain could be replaced by any functionally equivalent system, such as a giant lookup table or the population of China communicating with radios and satellites (Block 2006).

15 We have an intuition that we would notice if, for example, the implantation of the chip removed half of our visual consciousness. But according to the premises of the experiment, our behaviour would be identical, so nothing in our thoughts or speech would reflect this change in consciousness. If the implanted chip did affect our consciousness we would not be cognitively aware of the change and it would not affect our ability to perceive and respond to the world. We would be like people with anosognosia (see Chapter 4, Footnote 27), with the difference that our sight and bodies would be working perfectly, so no external observer could detect the change in our consciousness.

16 It might be argued that neurons die all the time, so surely replacing *one* neuron with silicon should not affect our assumption that the brain is a platinum standard? And so on with two neurons, three neurons, until the entire brain has been replaced. Chalmers' (1995a) fading and dancing qualia argument proceeds along these lines. One problem with this

argument is that it is based on the invalid assumption that we can imagine the relationship between consciousness and the brain (see Chapter 3). Another problem is that the brain can be sensitive to individual spikes, so the replacement of individual neurons could affect its consciousness. For example, a single neuron could individually encode an abstract concept or make a significant contribution to a population code. If this neuron was part of a CC set, then its replacement with a silicon chip could alter the associated conscious state.

17 The assumption that brains with implanted chips are conscious is equivalent to the assumption that functionalism is true. This brings in all of the problems with computation and information theories of consciousness that are discussed in Chapters 7 and 8.

18 Popper (2002, pp. 279-80).

19 Tononi (2008, p. 217). I also could have quoted Dehaene: 'Only mathematical theory can explain how the mental reduces to the neural. Neuroscience needs a series of bridging laws, analogous to the Maxwell-Boltzmann theory of gases, that connect one domain with the other. This is no easy task: the "condensed matter" of the brain is perhaps the most complex object on earth. Unlike the simple structure of a gas, a model of the brain will require many nested levels of explanation. In a dizzying arrangement of Russian dolls, cognition arises from a sophisticated arrangement of mental routines or processors, each implemented by circuits distributed across the brain, themselves made up of dozens of cell types. Even a single neuron, with its tens of thousands of synapses, is a universe of trafficking molecules that will provide modelling work for centuries.' (Dehaene 2014, p. 163).

20 I have discussed the need for c-theories elsewhere (Gamez 2012b).

21 The search for c-theories is closely related to the attempt to discover the relationship between brain activity and behaviour. Computational methods could also be used to study this relationship (see Section 5.6). However, c-theories might be based on non-neural structures in the brain, such as novel materials, haemoglobin and electromagnetic waves (see Section 6.2 and Section 6.3), that would not be required by theories that describe the relationship between neuron activity and external behaviour.

22 'Mathematics' should be interpreted in a broad sense that includes computer algorithms.

23 This example and its intensity values are purely illustrative. More work needs to be done on the conversion of c-reports into c-descriptions that

record the intensity of different aspects of conscious experience. This could draw on previous work in psychophysics—for example, Gescheider (1997).

24 This is an unashamedly Popperian approach to the science of consciousness. Some would argue that Popper (2002) presents an outmoded account of the philosophy of science, which should be replaced by Kuhn (1962) at least, or perhaps Feyerabend (1975) or Latour (1987). Some of these later 'relativist' 'constructivist' 'postmodern' accounts reject the possibility of scientific progress altogether. However, if we are attempting to understand how a science of consciousness can be developed, we need a model of what science is. And I would argue that Popper provides a carefully thought out and convincing account of what good scientific practice should be. Other philosophies of science can be used to interpret the science of consciousness, but many of them are considerably less useful as guiding principles than Popper—how (or why) would we develop a science of consciousness based on Feyerabend or Latour?

25 C-theories describing brute regularities have some similarity with Chalmers' psychophysical laws: 'Where we have new fundamental properties, we also have new fundamental laws. Here the fundamental laws will be *psychophysical laws*, specifying how phenomenal (or protophenomenal) properties depend on physical properties. These laws will not interfere with physical laws; physical laws already form a closed system. Instead they will be *supervenience laws*, telling us how experience arises from physical processes. We have seen that the dependence of experience on the physical cannot be derived from physical laws, so any final theory must include laws of this variety.' (Chalmers 1996a, p. 127). However, this book suspends judgment about some of the metaphysical substance-based theories, and the relationship between c-descriptions and p-descriptions is symmetrical, not a causal relationship in which consciousness *arises* from physical processes.

26 For example, Tononi's (2008) information integration theory is based on his first-person observations about the differentiation and integration of consciousness.

27 Humphreys (2004, p. 90).

28 There is also a more general question about whether one human brain can fully understand another—one might think that a brain could only be understood by a larger and more complicated system. This issue can potentially be addressed by using the world as external memory (Clark 2008; O'Regan 1992). This would only work if our understanding of the brain can be broken down into interrelated modules. For example, we

could develop a detailed understanding of how a neuron works, write it down, and then work on a different aspect of the problem, until we had written down everything about the brain. Although the final solution could not be comprehended by a single brain all at once, one or more brains could check the validity of each part and the links between them.

29   A substantial amount of research has been carried out on the use of computers for scientific discovery (Dzeroski and Todorovski 2007). Robotic systems have been developed that can carry out experiments automatically (Sparkes et al. 2010), and there has been research on the automatic discovery of differential equations that describe the behaviour of dynamic physical systems (Schmidt and Lipson 2009). This work suggests how consciousness could be scientifically studied in the future.

30   For example, Billeh et al. (2014) have developed a way of identifying functional circuits in recordings of spiking activity from hundreds of neurons. Using this approach it might be possible to develop a way of describing brain activity in terms of interacting circuits, which could be identified automatically by a computer.

31   The Blue Brain Project has developed detailed models of a cortical column (Markram 2006) and this work is being continued on a larger scale in the Human Brain Project (www.humanbrainproject.eu). Larger, less detailed models have also been built of human and animal brains (Ananthanarayanan et al. 2009; Izhikevich and Edelman 2008). The feasibility of scanning and simulating a human brain is discussed in Chapter 11, Footnote 14. None of the current models generates behaviour that is similar to c-reports and most of them do not include non-neural components of the brain, such as glia.

32   The 'c-reports' of a simulated brain could not be used to measure its consciousness because a neural simulation is not a platinum standard system.

33   Simulations are very different from real brains, so this would primarily be a test of the methodology. However, this type of work might lead to c-theories that could be tested on real brains.

34   This methodology could also be used to solve the more general problem of the relationship between an organism's brain activity and all of its behaviour (both conscious and non-conscious). Once the behaviour had been formally described, computers could be used to discover relationships between the brain activity and behaviour. This approach could be prototyped on a very simple system, such as a simulated C. elegans.

6. Physical Theories of Consciousness

1. The pattern/material distinction captures a useful way of speaking about the physical world at different spatial scales. However, one can also argue that elementary wave-particles are the only material and all other 'materials' are patterns in elementary wave-particles. Physical c-theories can be expressed using either interpretation of the pattern/material distinction.

2. A neuron's distance from the North Pole is a property of the neuron and the North Pole combined—it changes when the location of the North Pole changes. The brain has intrinsic properties that enable it to reflect particular frequencies of electromagnetic waves. The set of electromagnetic waves that is actually reflected depends on the brain's properties and on the properties of the waves. If electromagnetic waves altered their nature, the brain's reflectance of electromagnetic waves would change.

3. I have discussed this issue in more detail in a paper that distinguishes between type A correlates of consciousness that meet constraint C4, and type B correlates of consciousness that do not (Gamez 2014c).

4. A quantum theory of consciousness has been put forward by Hameroff and Penrose (1996). Electromagnetic theories of consciousness have been put forward by Pockett (2000) and Macfadden (2002).

5. The potential connection between consciousness and a global workspace was first elaborated by Baars (1988). A number of computational and neural models of a global workspace have been built (Franklin 2003; Gamez et al. 2013; Shanahan 2008; Zylberberg et al. 2010) and a substantial amount of research has been done on the possibility that a global workspace might be implemented in the brain (Dehaene and Changeux 2011).

6. For example, Bartfield et al. (2015) and Godwin et al. (2015) describe functional connectivity patterns that are potentially linked to consciousness.

7. See Gamez (2014b).

8. See Tononi and Sporns (2003), Balduzzi and Tononi (2008) and Oizumi et al. (2014). In a physical c-theory Tononi's algorithms would connect patterns in a particular material to a conscious state. This relationship would only hold for a specific material—the same patterns in a different material would not be linked to consciousness. This is distinct from the use of Tononi's algorithms to identify information patterns that are linked to consciousness, which is discussed in the next chapter. Liveliness (Gamez and Aleksander 2011), causal density (Seth et al. 2006) and Casali

et al.'s (2013) perturbational complexity index can also be re-interpreted as descriptions of patterns in materials that might be linked to consciousness.

9   A formal description of biological structures is required if CC sets contain biological materials and we want to make predictions or deductions (see Chapter 9) about the consciousness of non-biological systems. For example, a formal description of neurons could help us to decide whether a robot controlled by artificial neurons is conscious.

10 Lucretius' (2007) theory about the soul is similar to this view. He claims that the soul (a combination of spirit [*anima*, the vital principle] and mind [*animus*, the intellect]) is a minute particle:

> The nature of the mind and spirit is such it must consist
> Of stuff composed of seeds that are so negligibly small,
> Subtracted from the flesh, they don't affect the weight at all.
> Nor should we think this substance is composed of one thing, neat,
> For from the dying there escapes a slight *breath* mixed with *heat*,
> While heat, in turn, must carry *air* along with it; for there
> Is never any heat that is not also mixed with air,
> Because heat's substance, being loose in texture, has to leave
> Space for many seeds of air to travel through its weave.
> This demonstrates the nature of the mind's at least threefold –
> Even so, these three together aren't enough, all told,
> To generate sensation, since the mind rejects the notion
> Any of these is able to produce sense-giving motion,
> Or the thoughts the mind itself turns over. And so to these same
> Three elements, we have to add a fourth that has no name.
> There is nothing nimbler than this element at all –
> Nothing is as fine as this is, or as smooth or small.
> It's this that first distributes motions through the frame that lead
> To sense, since this is first to bestir, composed of minute seed.
>
> Lucretius (2007, pp 78-9)

11 At most people have invoked the known properties of quantum mechanics, which are unlikely to play much of a role (Wilson 1993).

12 Novel materials will not help us to imagine the relationship between consciousness and the physical world. If they are similar to the rest of the physical world, then they will be invisible, and we will be unable to make an imaginative transition from the invisible novel material to conscious experiences (see Section 3.3). If the novel material is more like a spark of consciousness embedded in matter, then we will be able to imagine the material, but we will find it difficult to imagine how it relates to other conscious experiences (see Section 3.4).

13   To make this assumption work it will be necessary to find a way of comparing the strength of patterns in different materials. For example, how can you compare the strength of electromagnetic field patterns (measured in volts) with blood flow patterns (measured in cm/s)?

7. Information Theories of Consciousness

1   Tononi (2008, p. 237).

2   Floridi (2010, p. 1).

3   Floridi uses 'dedomena' to describe differences in the physical world that exist independently of us: 'Dedomena are [...] pure data or proto-epistemic data, that is, data before they are epistemically interpreted. As "fractures in the fabric of being" they can only be posited as an external anchor of our information, for dedomena are never accessed or elaborated independently of a level of abstraction [...] They can be reconstructed as ontological requirements, like Kant's noumena or Locke's substance: they are not epistemically experienced but their presence is empirically inferred from (and required by) experience. Of course, no example can be provided, but dedomena are whatever lack of uniformity in the world is the source of (what looks to information systems like us as) as data, e.g. a red light against a dark background.' (Floridi 2009, pp. 17-8).

4   This notion of an interface is based on Floridi's level of abstraction: '[...] data are never accessed and elaborated (by an information agent) independently of a *level of abstraction* (LoA) [...]. A LoA is a specific set of typed variables, intuitively representable as an interface, which establishes the scope and type of data that will be available as a resource for the generation of information.' (Floridi 2009, p. 37). Floridi (2008) describes levels of abstraction in detail.

5   I can also extract the text of *Madame Bovary* from the DRAM voltages by defining a mapping between 01110010011001010110010 and the complete text of *Madame Bovary*.

6   A time-indexed interface uses a combination of time and the system's state to extract information. Suppose an elementary wave-particle shifts between two states: you can interpret the appearance of state 1 at time 1 as 'r', the appearance of state 1 at time 3 as 'e', and so on.

7   The notion of a custom interface is inspired by discussions about whether physical systems implement finite state automata (Bishop 2002; Bishop 2009; Chalmers 1996b; Chrisley 1995b; Putnam 1988).

8   According to Floridi's (2009) formulation of the general definition of information, σ is an instance of information, understood as semantic content, if and only if: 1) σ consists of n data, 2) the data are well formed, 3) the well-formed data are meaningful. My earlier work used this distinction to analyze Tononi's information integration theory of consciousness (Gamez 2011; Gamez 2016). I am indebted to Laurence Hayes for helping me to see that the data/information distinction is unworkable.

9   See Shannon (1948). Tononi's (2004; 2008) information integration theory of consciousness is based on this interpretation of information.

10 There are several versions of Tononi's information integration theory of consciousness (Balduzzi and Tononi 2008; Oizumi et al. 2014; Tononi 2004). Tononi (2008) gives a good overview and his book offers a simple introduction without the mathematical treatment (Tononi 2012). Experimental work on the information integration theory of consciousness has been carried out by Lee et al. (2009), Massimini et al. (2009), Ferrarelli et al. (2010), and Casali et al. (2013).

11 Barrett (2014) suggests how Tononi's information integration theory of consciousness can be interpreted as a physical c-theory.

12 Tononi (2008) suggests that his algorithm could be applied to all possible levels of the brain—the level at which it reaches a maximum would be the one that is correlated with consciousness.

13 It might be objected that if information is subjective, then surely it must be present in the brain? Where else can subjective things be? However, the neural mechanisms (and electromagnetic fields etc.) that are active when our brain applies an interface to a physical system are purely physical processes—they do not have special informational properties that are absent from the rest of the physical world. These physical mechanisms are associated with bubbles of experience in which colours, abstract concepts and 1s and 0s appear. The presence of 1s and 0s in consciousness does not prove that there are 1s and 0s in our physical brains, any more than the presence of red in consciousness proves that there is red in our physical brains.

14 It could be argued that an observer has to exchange physically conserved quantities with a system to read its state. This issue can be avoided if the observer applies an interface to emissions from the system, such as light patterns from a screen.

15 An information c-theorist might argue that a material *implementation* of information has e-causal powers. The material holds the pattern of

information and this pattern affects future states of the physical system. However, in this case the material must be considered to be implementing every possible information set that can be read from the system. Some of these are contradictory or have no relationship with each other. It is implausible to claim that a potentially infinite collection of disparate information sets are present in the material and lead to its state transitions.

16 The system could also extract information about an earlier state of itself.

17 I have discussed this experiment elsewhere (Gamez 2016). It would only work if the problems with custom-designed interfaces could be addressed.

18 For example, Tononi's information integration algorithms might be able to identify neuron firing patterns that are linked to consciousness. If this was a physical c-theory, these patterns would not be associated with consciousness when they occurred in other materials.

8. Computation Theories of Consciousness

1 Kentridge (1994, p. 442).

2 The MONIAC was a water computer that was developed by Bill Phillips to model the UK economy.

3 The solution is not necessarily optimal. A video can be found here: www.youtube.com/watch?v=dAyDi1aa40E

4 This discussion sets aside issues about time slicing and parallel processing, which do not affect the central argument. When a general-purpose computer is parallel processing it is executing multiple special-purpose computers simultaneously. When a general-purpose computer is time slicing it is working as a particular special-purpose computer for short periods of time.

5 This used to be a common practice—see Grier (2005).

6 Some of the computations that might be members of CC sets are discussed by Cleeremans (2005). Jackendof (1987) and Bor (2012) set out computational c-theories and Metzinger (2003) gives informational-computational interpretations of his constraints on conscious experience.

7 Computation c-theories are similar to a philosophical position known as functionalism, which claims that functions are the sole members of CC sets. Putnam (1975) was one of the key advocates of this position, which he later abandoned; Shagrir (2005) gives a good overview.

8   Standard orreries do not include the date—this could easily be added.

9   Strictly speaking, information cannot be stored *in* the physical world. To store information we make changes to the physical world that are defined by an interface. At a later point in time we access the same part of the physical world through the same interface and the information reappears.

10 Horsman et al. (2014) give a good description of this interpretation of computing.

11 This is based on the first example in Section 7.1.

12 This is true of any physical system because an interface can be custom-designed to extract a given sequence of information states from any sequence of physical states (Putnam 1988). The simplest method would be to map unique states onto the required information, or a clock could be used to handle repeated physical states. This mapping can only be done retrospectively unless one has a good predictive model of how the system's physical states will change.

13 For example, it has been claimed that everything is a cellular automata (Wolfram 2002; Zuse 1970) or that physical reality arises from the posing of yes-no questions—Wheeler's (1990) 'It from bit' hypothesis.

14 This theory of implementation will have to map spatiotemporal physical structures onto computations. It cannot be based on information, which only exists relative to a human-defined interface.

15 Putnam (1988) and Bishop (2002; 2009) discuss the problems with finite state automata.

16 I have described the problems with combinatorial state automata in a paper (Gamez 2014a) that raises more general problems with theories of implementation.

17 Piccinini (2007) puts forward a theory of implementation based on string processing.

18 Theories of implementation based on cellular automata have been put forward by Zuse (1970), Wolfram (2002) and Schule (2014). Piccinini (2012) gives a good overview of different theories of implementation and their problems.

19 Computational and functional concepts can be useful ways of describing physical correlates of consciousness. Suppose someone claims that consciousness is linked to a global workspace in the brain (Baars 1988). If this was interpreted as a physical c-theory, the global workspace would

just be a convenient way of describing a pattern in spiking neurons. The global workspace would not form a CC set by itself and it would not be correlated with consciousness if it was implemented in a different physical system.

9. Predictions and Deductions about Consciousness

1. Popper (2002, p. 18).

2. Research on change blindness suggests that we cannot accurately recall earlier conscious states. See, for example, Simons and Rensink (2005).

3. I am indebted to Ron Chrisley for this suggestion. To convert a c-description into a virtual reality file (for example, an X3D XML file) it is necessary to model the connection between virtual environments and states of consciousness. This will have to take the limited resolution of the senses and the active nature of the visual system into account. There will also be a one-to-many mapping between a c-description of a conscious state and virtual environments that could produce this state. This method of validating consciousness is limited to online consciousness that is evoked by sensory input. Many aspects of consciousness, such as body states and emotions, are difficult to control with virtual reality technology.

4. This is similar to the approach that is used in experiments on brain reading. For example, in one set of experiments by Nishimoto et al. (2011) the subjects watched a video while their brains were measured. The scientists used this data to build a model of the spatiotemporal structures in their brains that were activated by the video. This model was then used to reconstruct the video that the subjects were watching from their brain activity. The subjects could compare the consciousness that they had when they watched the reconstructed videos with the consciousness that they had when they watched the original video.

5. This testing method is easier if there is a 1:1 relationship between conscious states and physical states. Otherwise, the c-theory will map each conscious state onto multiple potential physical states.

6. There has been much speculation about whether a head remains conscious after it has been cut off. Dash (2011) discusses some of the early experiments that were carried out on humans. One study on rats suggests that they retain consciousness for several seconds after decapitation, and a wave of potentially conscious activity occurs approximately 50 seconds later (van Rijn et al. 2011). The EEG traces of dying humans show a similar pattern on a longer time scale (Chawla et al. 2009).

7   This is the classic problem raised by Nagel (1974) about what it is like to be a bat. From the perspective of this book, this is not a problem with the irreducibility of subjective experience, but with our limited ability to transform our bubble of experience into a different bubble of experience. There is no philosophical problem about deducing a c-description of a bat's consciousness from a p-description of its physical state—just a problem with our ability to imaginatively comprehend the c-description we have generated.

8   One potential solution would be to create virtual reality environments that enable us to experience aspects of a bat's consciousness. Alternatively Chrisley (1995a) has suggested how we could use robotic systems to specify the non-conceptual contents of a bat's consciousness. This approach has been demonstrated by Chrisley and Parthemore (2007), who used a SEER-3 robot to specify the non-conceptual content of a model of perception based on O'Regan and Noë's (2001) sensorimotor theory.

9   This will only be possible if we have a flexible and general c-description format (see Section 4.9).

10   For example, deductions could help us to breed or genetically engineer food animals that are not conscious.

11   D could also be a constant pattern or a partially correlated pattern (see Section 6.4).

12   The concept of similar physical contexts needs to be worked out in detail. Normally functioning adult human brains have a great deal of variability in their patterns and materials, so a statistical definition of the normal variability in their patterns and materials is required to precisely define a physical context.

13   Conservative deductions could be made about unreportable consciousness in a platinum standard system during an experiment on consciousness. Suppose a brain contains two identical structures, one connected to c-reports and one not. If these structures were always present together, pilot studies would identify their union as the correlate. However, if the structure that was disconnected from c-reports came and went intermittently, then we could exclude it as the correlate. At a later point in time when both structures were present we might deduce that there are two consciousnesses in the platinum standard system, only one of which is reportable. This would violate assumptions A1, A2 and A6. However, we would still need A1, A2 and A6 to identify the correlate that was used to make the deduction. An example of this type of reasoning can be found in Lamme (2006; 2010), who uses paradigmatic cases of reportable

consciousness to establish the link between consciousness and recurrent processing, and then makes inferences about the presence of inaccessible phenomenal consciousness.

14 In previous work I proposed that indeterminacy envelopes could be used to make liberal deductions about consciousness (Gamez 2012a). I have replaced this with the framework that is described in this book.

15 The conservative/liberal distinction is based on a binary opposition between similar and different physical contexts. The differences between physical contexts could also be expressed as a continuous value, which would correspond to the degree of liberality of the deduction.

10. Modification and Enhancement of Consciousness

1 James (1985, p. 388).

2 A fused consciousness would be separately created in my brain and your brain—there would not be any merging of our actual consciousnesses.

3 As explained in Section 2.5, the overall level of consciousness is something like the average level of intensity of the properties and objects in a bubble of experience. This can be reduced with anaesthetics, such as propofol, or by a blow to the head. Chemicals, such as caffeine or LSD, can increase the overall level of intensity. It can also be increased by emotionally intense situations, such as a car crash.

4 Sensory input is the main method that we use to change the contents of our consciousness. If I want an elephant in my bubble of experience, then I go to the zoo and look at an elephant. Hallucinogenic drugs have a strong effect on contents and we have some control over contents in lucid dreams and imagination.

5 This type of experience is well documented (Crookall 1972) and it can be induced through body trauma, mental exercises (Harary and Weintraub 1989; Ophiel 1970), or chemicals, such as ketamine (Wilkins et al. 2011). Out-of-body experiences can also occur in brain-damaged patients (Blanke and Arzy 2005) and psychology experiments can induce the illusion that part or all of our bodies are in a different location (Ehrsson 2007). There is no compelling evidence to suggest that people who are having an out-of-body experience can report information about the physical world that has not been obtained through the senses of their physical body (Alvarado 1982; Blackmore 2010).

6   Sensory manipulation can alter the perceived size of our body in relation to our environment (van der Hoort et al. 2011). Muscimol (found in the mushroom Amanita Muscaria) is reported to be capable of this.

7   Castaneda (1968) describes how he used a combination of mental control and hallucinogens to transform his conscious experience of his body into a crow (the truth of his account has been disputed). Phantom limbs demonstrate that our experiences of our bodies are linked to brain activity and are distinct from our actual physical bodies (Gamez 2007, pp. 57-60; Melzack 1990; Melzack 1992). This suggests that the shape of our consciously experienced bodies can be altered by modifying our brains.

8   Sensory input, such as looking at fearful or beloved objects, changes our emotional states. Chemicals, such as cocaine or Prozac, alter the intensity of our emotional states on short or long time scales.

9   The current size of our bubbles of experience is probably linked to the size of our brains. More brain tissue is likely to be required to expand our bubbles of experience without loss of resolution.

10 Chakravarthi and VanRullen (2012) describe experimental evidence for the discrete nature of conscious perception. VanRullen and Koch (2003) have a more general discussion of this issue. There are well-documented examples of people with expanded long term memories—a condition known as hyperthymesia (Parker et al. 2006). Borges' (1970) Funes the Memorious is a fictional example.

11 Animals with different senses (for example, bats and fish) are likely to have different sensations. I have suggested elsewhere that conscious sensations might be linked to the neural patterns caused by sensory input, and that our conscious perception of a three-dimensional world could be linked to a combination of sensory and sensorimotor patterns (Gamez 2014b). If this is the case, then a novel sensory pattern would be associated with a novel sensation.

   Attempts have been made to create novel sensations. For example, the feelSpace belt gives subjects information about the location of North (Nagel et al. 2005) and magnetic fingertip implants enable people to feel magnetic fields. However, it is not clear whether these devices give people new conscious sensations. This is probably because the novel sensory input is processed through the existing senses, instead of being directly fed into the cortex.

12 We understand the link between changes in sensory input and changes in consciousness, but we do not understand *how* changes in sensory input lead to changes in the brain that are associated with an altered

consciousness. The same is true for imagination and the ingestion of consciousness-modifying chemicals.

13 We would be unlikely to remember some of the conscious states that could be induced in us. Episodic memories regenerate earlier states of our brains. This might not be possible if a CC set is not the consequence of the brain's own activity.

14 This technology is dramatized in the 1995 film *Strange Days*. It is different from a virtual reality system, which only mimics the sensory input produced by an environment and has little effect on our conscious experience of our body.

15 Patterns in electromagnetic fields, glia and blood can be indirectly manipulated by changing the neuron activity.

16 See Legon et al. (2014).

17 For example, electrodes have been used to modify the memories of mice (de Lavilleon et al. 2015; Ramirez et al. 2013).

18 For example, see Nikolenko et al. (2007). Electrodes and optogenetics are unlikely to be able to increase a neuron's firing rate beyond a certain point because of metabolic constraints.

19 Chen et al. (2015) have developed a method for brain stimulation that uses magnetic nanoparticles. Seo et al. (2013) have outlined a design for a wireless brain interface that uses thousands of biologically neutral microsensors to convert electrical signals into ultrasound that can be read outside the brain. This could potentially be extended to deliver signals to the brain.

20 This might be required if we want to expand our spatial and temporal consciousness.

21 Whether a synthetic neuron is a valid member of a CC set will depend on how neurons are p-described (see Section 5.1).

22 Additional neurons would only alter consciousness if CC sets consist of neuron activity patterns or if the additional neurons altered CC sets in some other way—for example, by changing the electromagnetic fields.

23 For example, Yin et al. (2013) have developed a wireless electrode interface that is implanted below the skin.

24 A further problem is that invasive technologies are only allowed under very specific conditions on human subjects. This may change if the safety

of these techniques is demonstrated and there is demand or demonstrable benefits. A workable technology will also be appropriated by the public at large regardless of the safety issues or legal constraints. For example, you can buy tDCS kits on the Internet.

25 Huxley (1965, pp. 71-2).

11. Machine Consciousness

1 Metzinger (2003, p. 618).

2 Searle (1980, p. 424).

3 Machine consciousness is also called artificial consciousness. I have presented a version of these types of machine consciousness elsewhere (Gamez 2008b). They overlap with Seth's (2009) distinction between strong and weak artificial consciousness and have some similarity with Searle's (1980) distinction between strong and weak artificial intelligence. More information about previous work on machine consciousness is given by Holland (2003), Chella and Manzotti (2007), Gamez (2008b) and Reggia (2013). The *International Journal of Machine Consciousness* has published many papers on this topic.

4 This is sometimes known as artificial general intelligence (AGI).

5 Arrabales' (2010) ConsScale ranks systems according to their MC1 consciousness. The Turing test and Harnad's (1994) variations of it are designed to test whether a system exhibits the full spectrum of human behaviour. There has been a large amount of work on MC1 systems — virtually any computer capable of perception and learning can be interpreted as a MC1 machine.

6 For example, computer models of global workspaces have been built (Franklin 2003; Gamez et al. 2013; Shanahan 2008; Zylberberg et al. 2010).

7 A number of people have used internal models that are updated with sensory data to control robots—for example, Chella, Liotta and Macaluso (2007). Computer models have also been built of imagination (Gravato Marques and Holland 2009) and of sensorimotor theories of consciousness (Chrisley and Parthemore 2007).

8 For example, I have collaborated on a global workspace model implemented in spiking neurons (MC2), which produced human-like behaviour (MC1) in the Unreal Tournament computer game (Gamez et al. 2013).

9   Digital computers that are simulating neurons produce very different electromagnetic fields from biological neurons. Neuromorphic chips use the flow of electrons to model the movement of ions in biological neurons (Indiveri et al. 2011). This type of chip is more likely to produce similar electromagnetic fields to biological neurons.

10   Neurons cultured in a Petri dish have been used to control a virtual animal (Demarse et al. 2001) and a robot (Warwick et al. 2010).

11   I have carried out preliminary experiments that illustrate how deductions can be made about the consciousness of an artificial system (Gamez 2008a; Gamez 2010).

12   The implantation of non-conscious chips that modify CC sets in the brain is covered in Section 10.4. The implantation of chips to study the relationship between consciousness and the physical world is covered in Section 5.4.

13   See Footnote 9.

14   The connections between neurons have traditionally been identified by the laborious method of injecting tracers (Zingg et al. 2014). More promising techniques are starting to emerge that might be able to automatically scan dead human brains. For example, knife-edge scanning microscopes can automatically slice and photograph brain tissue, which enables some of the neurons and connections to be discovered (Mayerich et al. 2008). However, this technique can only identify a limited number of neurons and it cannot reveal the direction of connections. A more promising direction is the automation of electron microscopy to mill and scan blocks of brain tissue. With further development this approach might be able to identify all of the neurons and connections in an adult human brain (Knott et al. 2008). There has also been research into techniques for making dead tissue transparent, which could help us to map the neurons and connections (Yang et al. 2014).

When the neurons and connections have been identified the next challenge is to simulate them on a computer. The adult human brain has approximately 100 billion neurons and 10^{14} connections. Networks with around a billion point neurons and 10^{13} connections have been simulated much slower than real time (Ananthanarayanan et al. 2009) and the SpiNNaker project is working towards the goal of simulating a billion neurons in real time (Furber and Temple 2007; Rast et al. 2011). One critical question is how much of the neurons' structure will need to be simulated to reproduce the brain's large-scale behaviour. If it is a large amount, then it is going to take a lot longer to reach the point at which it can be done in real time. It is also unlikely that we will be able to realize CC sets in artificial systems by simulating neurons. Neuromorphic chips have a greater chance of reproducing the electromagnetic fields of biological neurons, and it should soon be possible to run a million of these in real time (Benjamin et al. 2014).

15   This possibility has been dramatized in the 2014 film *Transcendence*. Carbon Copies is a non-profit organization that promotes the scanning and uploading of brains (www.carboncopies.org).

16   Imagine a scenario in which an artificial implant (made from appropriate materials) was added to your brain and the pattern associated with 1% of your consciousness was on the implant, with the rest of the pattern in your brain. The proportion on the implant could be progressively increased until the entire pattern was on the implant. This would still be a process of copying in which the original is progressively lost. At the beginning of this process there would be a consciousness associated with your brain and no consciousness associated with the implant. At the end, a copy of your consciousness would be associated with the implant and your brain would have lost consciousness. In the intermediate cases, there would be some of the original consciousness and some of the copy. This type of gradual replacement of materials happens all the time in the brain as it exchanges atoms with its surroundings. So in practice we cannot avoid the gradual replacement of our consciousness as the material in our brains changes. At best we can minimize such changes—for example, by not agreeing to copies of our consciousness that destroy the original.

The identity of bubbles of experience over time is similar to the identity of physical objects over time. Some changes to a motorbike have minimal impact on our sense of its continuity—for example, replacing the spark plugs. Other changes, such as swapping the chassis or making an atom-for-atom copy have a bigger effect. When large changes are made, I might prefer the original because of its history—it is my motorbike, the motorbike that I rode around the world, and so on. Other people might not care whether they have the original or an atom-for-atom copy. In a similar way, some people might believe that the arrangement of their bubble of experience is what is important—these people are happy as long as this arrangement exists somewhere. This is equivalent to the atom-for-atom copy of the motorbike. Other people prefer the consciousness that is linked to their brain and believe that they will die when this consciousness ceases, regardless of whether a copy has been made somewhere. This is equivalent to preferring the original motorbike with none of its parts replaced.

17   Kaczynski (1996) and Joy (2000) believe that we will increasingly pass responsibility to intelligent machines until we are unable to do without

them—in the same way that we are increasingly unable to live without the Internet today. This might eventually leave us at the mercy of super-intelligent machines who could use their power against us. Kaczynski killed three people and injured twenty-three others to raise awareness of this issue.

18   Most people are concerned about machines that behave like conscious human beings, so I am setting aside the possibility that machines could produce non-conscious external behaviour that threatens humanity.

19   We might be able to produce a MC1 machine by scanning a human brain and simulating it on a computer (see Footnote 14). This would not be any more intelligent than us or any more of a threat to humanity than an intelligent human. However, it would be easier to understand and improve than a human brain, so it could be the starting point for more advanced forms of intelligence. In the medium term it might become possible to run simulations of brains faster than biological brains and to run multiple simulated brains in parallel. Deep learning is another promising method for producing MC1 machines. For example, Mnih et al. (2015) used deep reinforcement learning to train a neural network to play 1980s video games with human-level performance.

20   For example, machines would have to have human level intelligence; they would have to be capable of powering and maintaining themselves for long periods of time; military computers would have to be connected to the Internet and inadequately defended against hackers; etc. Machines would also become a threat if they became good at manipulating human behaviour.

21   Chalmers (2010) has a good discussion of the singularity. Eden et al. (2012) have edited a collection of papers on this topic.

22   If we had a mathematical way of measuring intelligence, then genetic algorithms could be used to create systems with a high value of this measure. A number of universal intelligence measures have been put forward (Hernández-Orallo and Dowe 2010; Hibbard 2011; Legg and Hutter 2007), but I am not aware of any that would be suitable for this task.

23   The construction of a system that can produce something that is more intelligent than itself is extremely challenging. It will not happen by accident, but through many years of laborious trial and error. Papers will be published on prototypes, there will be early versions that are partly functional, and so on. Only when the technology has been tried in many different ways is there any possibility that it could create a super-intelligence.

24 One of the most dangerous computer errors was a malfunction in the Soviet nuclear early warning system in 1983, which almost led to a third world war. Asimov (1952) dramatizes some of the problems with malfunctioning intelligent machines.

25 Sloman (2006).

26 This position is put forward by Moravec (1988) and Asimov (1952).

27 For example, murder entails the premature loss of the victim's consciousness and creates suffering in the bubbles of experience of the bereaved. On the other hand, switching off the life support of a coma patient is not generally considered wrong if the patient is not conscious and if they have no chance of regaining consciousness.

28 This might be the only way in which consciousness and our cultural traditions could survive the death of the sun in 5.4 billion years. It is unlikely that humans will be able to physically travel beyond our solar system to escape the dying sun. Machines can go much further because they can accelerate faster, feed on light and shut down for thousands of years while travelling.

29 Metzinger (2003, p. 621).

12. Conclusion

1 Popper (2002, p. 94).

2 Metzinger describes the current state of consciousness research as follows: 'The interdisciplinary project of consciousness research, now experiencing such an impressive renaissance with the turn of the century, faces two fundamental problems. First, there is yet no single, unified and *paradigmatic* theory of consciousness in existence which could serve as an object for constructive criticism and as a backdrop against which new attempts could be formulated. Consciousness research is still in a preparadigmatic stage. Second, there is no systematic and comprehensive catalogue of *explananda*. Although philosophers have done considerable work on the *analysanda*, the interdisciplinary community has nothing remotely resembling an agenda for research. We do not yet have a precisely formulated list of explanatory targets which could be used in the construction of systematic research programs.' (Metzinger 2003, pp. 116-7).

3 For example, whether consciousness is a non-physical substance, the hard problem, solipsism, zombies, colour inversion and the causal relationship between consciousness and the physical world.

4   In the standard version of dualism there is a bidirectional e-causal relationship between non-physical consciousness and the physical world. This is ruled out by assumption A5.

5   Assumptions A7-A9 are more pragmatic and might not be needed by the science of consciousness.

6   NeuroNexus sells electrodes that can record from 256 locations. It is working to expand this to 1,000 electrodes (Marx 2014).

7   See Ahrens and Keller (2013).

8   For example, Shanahan and Wildie (2012) have proposed a 'knotty centrality' measure that might be linked to consciousness.

9   An assumption that a mammalian brain is a platinum standard system will have much less impact on the science of consciousness than a similar assumption about an artificial system.

10 Wittgenstein (1969, remark 94).

11 For example, consider the assumption that all conscious states associated with a platinum standard system are linked to c-reports (A2). We could disprove this assumption if we could show that there are aspects of consciousness that are not accessible through c-reports. But since these aspects of consciousness cannot be accessed, we cannot prove that they do or do not exist.

12 This is the position of Berkeley (1957). Husserl (1960) developed his phenomenological program by suspending commitment to the reality of the physical world.

13 Hume (1993, p. 114).

14 At present the most mathematical theories of consciousness are information c-theories, which can be re-interpreted as physical c-theories. For example, Tononi's (2008) information integration theory of consciousness can be re-interpreted as a theory about the relationship between neuron activity and consciousness. Causal density (Seth et al. 2006), liveliness (Gamez and Aleksander 2011) and Casali et al.'s (2013) perturbational complexity index can also be reinterpreted as physical c-theories.

Bibliography

Ahrens, M. B. and Keller, P. J. (2013). Whole-brain functional imaging at cellular resolution using light-sheet microscopy. *Nature Methods* 10: 413–20. https://doi.org/10.1038/nmeth.2434

Alvarado, C. S. (1982). ESP during out-of-body experiences: a review of experimental studies. *Journal of Parapsychology* 46: 209–30.

Ananthanarayanan, R., Esser, S. K., Simon, H. D. and Modha, D. S. (2009). The cat is out of the bag: cortical simulations with 10^9 neurons, 10^{13} synapses. *Proceedings of Conference on High Performance Computing Networking, Storage and Analysis,* Portland, Oregon, ACM, pp. 1–12. https://doi.org/10.1145/1654059.1654124

Aronson, J. (1971a). The Legacy of Hume's Analysis of Causation. *Studies in History and Philosophy of Science* 2(2): 135–56. https://doi.org/10.1016/0039-3681(71)90028-8

Aronson, J. (1971b). On the Grammar of 'Cause'. *Synthese* 22(3/4): 441–30. https://doi.org/10.1007/bf00413436

Arrabales, R., Ledezma, A. and Sanchis, A. (2010). ConsScale: A Pragmatic Scale for Measuring the Level of Consciousness in Artificial Agents. *Journal of Consciousness Studies* 17(3–4): 131–64.

Aru, J., Bachmann, T., Singer, W. and Melloni, L. (2012). Distilling the neural correlates of consciousness. *Neuroscience & Biobehavioral Reviews* 36(2): 737–46. https://doi.org/10.1016/j.neubiorev.2011.12.003

Asimov, I. (1952). *I, Robot.* London: Grayson & Grayson.

Baars, B. J. (1988). *A Cognitive Theory of Consciousness.* Cambridge; New York: Cambridge University Press.

Balduzzi, D. and Tononi, G. (2008). Integrated information in discrete dynamical systems: motivation and theoretical framework. *PLoS Computational Biology* 4(6): e1000091. https://doi.org/10.1371/journal.pcbi.1000091

Balduzzi, D. and Tononi, G. (2009). Qualia: the geometry of integrated information. *PLoS Computational Biology* 5(8): e1000462. https://doi.org/10.1371/journal.pcbi.1000462

Barrett, A. B. (2014). An integration of integrated information theory with fundamental physics. *Frontiers in Psychology* 5: 63. https://doi.org/10.3389/fpsyg.2014.00063

Barttfeld, P., Uhrig, L., Sitt, J. D., Sigman, M., Jarraya, B. and Dehaene, S. (2015). Signature of consciousness in the dynamics of resting-state brain activity. *Proceedings of the National Academy of Sciences of the United States of America* 112(3): 887–92. https://doi.org/10.1073/pnas.1418031112

Benjamin, B. V., Gao, P., McQuinn, E., Choudhary, S., Chandrasekaran, A. R., Bussat, J.-M., Alvarez-Icaza, R., Arthur, J. V., Merolla, P. A. and Boahen, K. (2014). Neurogrid: A Mixed-Analog-Digital Multichip System for Large-Scale Neural Simulations. *Proceedings of the IEEE* 102(5): 699–716. https://doi.org/10.1109/jproc.2014.2313565

Berkeley, G. (1957). *A Treatise Concerning the Principles of Human Knowledge*. New York: Liberal Arts Press.

Bigelow, J., Ellis, B. and Pargetter, R. (1988). Forces. *Philosophy of Science* 55(4): 614–30. https://doi.org/10.1086/289464

Bigelow, J. and Pargetter, R. (1990). Metaphysics of Causation. *Erkenntnis* 33(1): 89–119. https://doi.org/10.1007/bf00634553

Billeh, Y. N., Schaub, M. T., Anastassiou, C. A., Barahona, M. and Koch, C. (2014). Revealing cell assemblies at multiple levels of granularity. *Journal of Neuroscience Methods* 236: 92–106. https://doi.org/10.1016/j.jneumeth.2014.08.011

Bishop, J. M. (2002). Counterfactuals cannot count: a rejoinder to David Chalmers. *Consciousness and Cognition* 11(4): 642–52. https://doi.org/10.1016/s1053-8100(02)00023-5

Bishop, J. M. (2009). A Cognitive Computation Fallacy? Cognition, Computations and Panpsychism. *Cognitive Computation* 1: 221–33. https://doi.org/10.1007/s12559-009-9019-6

Blackmore, S. J. (2010). *Consciousness: An Introduction*. London: Hodder Education.

Blake, R. (2001). A Primer on Binocular Rivalry, Including Current Controversies. *Brain and Mind* 2(1): 5–38. https://doi.org/10.1023/A:1017925416289

Blanke, O. and Arzy, S. (2005). The out-of-body experience: disturbed self-processing at the temporo-parietal junction. *Neuroscientist* 11(1): 16–24. https://doi.org/10.1177/1073858404270885

Block, N. (1995). On a Confusion about a Function of Consciousness. *Behavioral and Brain Sciences* 18(2): 227–47. https://doi.org/10.1017/s0140525x00038188

Block, N. (2006). Troubles with Functionalism. In *Theories of Mind: An Introductory Reader*, edited by M. Eckert. Maryland: Rowman & Littlefield, pp. 97–102.

Block, N. (2007). Consciousness, accessibility, and the mesh between psychology and neuroscience. *Behavioral and Brain Sciences* 30(5–6): 481–99. https://doi.org/10.1017/s0140525x07002786

Bor, D. (2012). *The Ravenous Brain*. New York: Basic Books.

Borges, J. L. (1970). *Labyrinths: Selected Stories and Other Writings.* London: Harmondsworth: Penguin.

Brown, J. R. and Fehige, Y. (2014) Thought Experiments. *The Stanford Encyclopedia of Philosophy (Fall 2014 Edition),* edited by. E. N. Zalta. http://plato.stanford.edu/archives/fall2014/entries/thought-experiment/

Casali, A. G., Gosseries, O., Rosanova, M., Boly, M., Sarasso, S., Casali, K. R., Casarotto, S., Bruno, M. A., Laureys, S., Tononi, G. and Massimini, M. (2013). A theoretically based index of consciousness independent of sensory processing and behavior. *Science Translational Medicine* 5(198): 198ra05. https://doi.org/10.1126/scitranslmed.3006294

Castaneda, C. (1968). *The teachings of Don Juan; a Yaqui way of knowledge.* Berkeley; Los Angeles: University of California Press.

Chakravarthi, R. and Vanrullen, R. (2012). Conscious updating is a rhythmic process. *Proceedings of the National Academy of Sciences of the United States of America* 109(26): 10599–604. https://doi.org/10.1073/pnas.1121622109

Chalmers, D. J. (1995a). Absent Qualia, Fading Qualia, Dancing Qualia. In *Conscious Experience,* edited by T. Metzinger. Thorverton: Imprint Academic, pp. 309–28.

Chalmers, D. J. (1995b). Facing Up to the Problem of Consciousness. *Journal of Consciousness Studies* 2(3): 200–19.

Chalmers, D. J. (1996a). *The Conscious Mind: In Search of a Fundamental Theory.* Oxford: Oxford University Press.

Chalmers, D. J. (1996b). Does a rock implement every finite-state automaton. *Synthese* 108: 309–33. https://doi.org/10.1007/bf00413692

Chalmers, D. J. (1998). On the Search for the Neural Correlates of Consciousness. In *Toward a Science of Consciousness II: The Second Tucson Discussions and Debates,* edited by S. Hameroff, A. Kaszniak and A. Scott. Cambridge, Massachusetts: MIT Press, pp. 219–29.

Chalmers, D. J. (2000). What Is a Neural Correlate of Consciousness? In *Neural Correlates of Consciousness,* edited by T. Metzinger. Cambridge, Massachusetts: MIT Press, pp. 17–39.

Chalmers, D. J. (2010). The Singularity: A Philosophical Analysis. *Journal of Consciousness Studies* 17(9–10): 7–65.

Chawla, L. S., Akst, S., Junker, C., Jacobs, B. and Seneff, M. G. (2009). Surges of Electroencephalogram Activity at the Time of Death: A Case Series. *Journal of Palliative Medicine* 12(12): 1095–100. https://doi.org/10.1089/jpm.2009.0159

Chella, A., Liotta, M. and Macaluso, I. (2007). CiceRobot: a cognitive robot for interactive museum tours. *Industrial Robot: An International Journal* 34(6): 503–11. https://doi.org/10.1108/01439910710832101

Chella, A. and Manzotti, R., (eds.) (2007). *Artificial Consciousness*. Exeter: Imprint Academic.

Chen, R., Romero, G., Christiansen, M. G., Mohr, A. and Anikeeva, P. (2015). Wireless magnetothermal deep brain stimulation. *Science* 347(6229): 1477–80. https://doi.org/10.1126/science.1261821

Chrisley, R. (1995a). Taking Embodiment Seriously: Nonconceptual Content and Robotics. In *Android Epistemology*, edited by K. M. Ford, C. Glymour and P. J. Hayes. Cambridge and London: AAAI Press/ The MIT Press.

Chrisley, R. (1995b). Why Everything Doesn't Realize Every Computation. *Minds and Machines* 4: 403–20. https://doi.org/10.1007/bf00974167

Chrisley, R. and Parthemore, J. (2007). Synthetic phenomenology—Exploiting embodiment to specify the non-conceptual content of visual experience. *Journal of Consciousness Studies* 14(7): 44–58.

Clark, A. (2008). *Supersizing the Mind: Embodiment, Action, and Cognitive Extension*. New York; Oxford: Oxford University Press.

Cleeremans, A. (2005). Computational correlates of consciousness. *Progress in Brain Research* 150: 81–98. https://doi.org/10.1016/s0079-6123(05)50007-4

Cohen, M. A. and Dennett, D. C. (2011). Consciousness cannot be separated from function. *Trends in Cognitive Sciences* 15(8): 358–64. https://doi.org/10.1016/j.tics.2011.06.008

Crookall, R. (1972). *Casebook of Astral Projection 546–746*. New York: University Books.

Damasio, A. (1999). *The Feeling of What Happens*. London: Vintage Books.

Dash, M. (2011). Some experiments with severed heads. https://allkindsofhistory.wordpress.com/2011/01/25/some-experiments-with-severed-heads

de Graaf, T. A., Hsieh, P. J. and Sack, A. T. (2012). The 'correlates' in neural correlates of consciousness. *Neuroscience & Biobehavioral Reviews* 36(1): 191–97. https://doi.org/10.1016/j.neubiorev.2011.05.012

de Lavilleon, G., Lacroix, M. M., Rondi-Reig, L. and Benchenane, K. (2015). Explicit memory creation during sleep demonstrates a causal role of place cells in navigation. *Nature Neuroscience* 18(4): 493–95. https://doi.org/10.1038/nn.3970

Dehaene, S. (2014). *Consciousness and the Brain: Deciphering how the Brain Codes our Thoughts*. New York: Penguin.

Dehaene, S. and Changeux, J. P. (2011). Experimental and theoretical approaches to conscious processing. *Neuron* 70(2): 200–27. https://doi.org/10.1016/j.neuron.2011.03.018

Dell'Acqua, R. and Grainger, J. (1999). Unconscious semantic priming from pictures. *Cognition* 73(1): B1–B15. https://doi.org/10.1016/s0010-0277(99)00049-9

Demarse, T. B., Wagenaar, D. A., Blau, A. W. and Potter, S. M. (2001). The Neurally Controlled Animat: Biological Brains Acting with Simulated Bodies. *Autonomous Robots* 11(3): 305–10. https://doi.org/10.1023/a:1012407611130

Dennett, D. C. (1992). *Consciousness Explained.* London: The Penguin Press.

Dowe, P. (2000). *Physical causation.* Cambridge: Cambridge University Press. https://doi.org/10.1017/cbo9780511570650

Dzeroski, S. and Todorovski, L., (eds.) (2007). *Computational Discovery of Scientific Knowledge: Introduction, Techniques, and Applications in Environmental and Life Sciences.* Berlin: Springer.

Eddington, A. S. S. (1928). *The Nature of the Physical World.* Cambridge: University Press.

Edelman, D. B. and Seth, A. K. (2009). Animal consciousness: a synthetic approach. *Trends in Neurosciences* 32(9): 476–84. https://doi.org/10.1016/j.tins.2009.05.008

Eden, A. H., Moor, J. H., Søraker, J. H. and Steinhart, E., (eds.) (2012). *Singularity Hypotheses: A Scientific and Philosophical Assessment.* Berlin: Springer.

Ehrsson, H. H. (2007). The experimental induction of out-of-body experiences. *Science* 317(5841): 1048. https://doi.org/10.1126/science.1142175

Fair, D. (1979). Causation and the Flow of Energy. *Erkenntnis* 14: 219–50. https://doi.org/10.1007/bf00174894

Feinberg, T. E. and Mallatt, J. (2013). The evolutionary and genetic origins of consciousness in the Cambrian Period over 500 million years ago. *Frontiers in Psychology* 4: 667. https://doi.org/10.3389/fpsyg.2013.00667

Fell, J., Elger, C. E. and Kurthen, M. (2004). Do neural correlates of consciousness cause conscious states? *Medical Hypotheses* 63: 367–69. https://doi.org/10.1016/j.mehy.2003.12.048

Ferrarelli, F., Massimini, M., Sarasso, S., Casali, A., Riedner, B. A., Angelini, G., Tononi, G. and Pearce, R. A. (2010). Breakdown in cortical effective connectivity during midazolam-induced loss of consciousness. *Proceedings of the National Academy of Sciences of the United States of America* 107(6): 2681–86. https://doi.org/10.1073/pnas.0913008107

Feyerabend, P. (1975). *Against Method: Outline of an Anarchistic Theory of Knowledge.* London: NLB.

Floridi, L. (2008). The Method of Levels of Abstraction. *Minds and Machines* 18(3): 303–29. https://doi.org/10.1007/s11023-008-9113-7

Floridi, L. (2009). Philosophical Conceptions of Information. *Lecture Notes in Computer Science* 5363: 13–53. https://doi.org/10.1007/978-3-642-00659-3_2

Floridi, L. (2010). *Information: A Very Short Introduction.* Oxford: Oxford University Press. https://doi.org/10.1093/actrade/9780199551378.001.0001

Franklin, S. (2003). IDA—A conscious artifact? *Journal of Consciousness Studies* 10(4–5): 47–66.

Friston, K. (1994). Functional and Effective Connectivity in Neuroimaging: A Synthesis. *Human Brain Mapping* 2: 56–78. https://doi.org/10.1002/hbm.460020107

Friston, K. J. (2011). Functional and effective connectivity: a review. *Brain Connectivity* 1(1): 13–36. https://doi.org/10.1089/brain.2011.0008

Froese, T., Gould, C. and Barrett, A. (2011). Re-Viewing from Within: A Commentary on First- and Second-Person Methods in the Science of Consciousness. *Constructivist Foundations* 6(2): 254–69.

Furber, S. and Temple, S. (2007). Neural systems engineering. *Journal of the Royal Society Interface* 4(13): 193–206. https://doi.org/10.1098/rsif.2006.0177

Galilei, G. (1957). *Discoveries and Opinions of Galileo. Including: The Starry Messenger (1610), Letter to the Grand Duchess Christina (1615); And Excerpts from: Letters on Sunspots (1613), The Assayer (1623).* Translated by S. Drake. New York: Doubleday & Co.

Gamez, D. (2006). The XML Approach to Synthetic Phenomenology. *Proceedings of AISB06 Symposium on Integrative Approaches to Machine Consciousness*, edited by R. Chrisley, R. Clowes and S. Torrance, Bristol, pp. 128–35.

Gamez, D. (2007). *What We Can Never Know: Blindspots in Philosophy and Science.* London: Continuum.

Gamez, D. (2008a). *The Development and Analysis of Conscious Machines.* Unpublished PhD Thesis, University of Essex.

Gamez, D. (2008b). Progress in machine consciousness. *Consciousness and Cognition* 17(3): 887–910. https://doi.org/10.1016/j.concog.2007.04.005

Gamez, D. (2009). The Potential for Consciousness of Artificial Systems. *International Journal of Machine Consciousness* 1(2): 213–23. https://doi.org/10.1142/s1793843009000190

Gamez, D. (2010). Information integration based predictions about the conscious states of a spiking neural network. *Consciousness and Cognition* 19(1): 294–310. https://doi.org/10.1016/j.concog.2009.11.001

Gamez, D. (2011). Information and Consciousness. *Etica & Politica / Ethics & Politics*, XIII(2): 215–34.

Gamez, D. (2012a). Empirically Grounded Claims about Consciousness in Computers. *International Journal of Machine Consciousness* 4(2): 421–38. https://doi.org/10.1142/s1793843012400240

Gamez, D. (2012b). From Baconian to Popperian Neuroscience. *Neural Systems and Circuits* 2(1): 2. https://doi.org/10.1186/2042-1001-2-2

Gamez, D. (2014a). Can we Prove that there are Computational Correlates of Consciousness in the Brain? *Journal of Cognitive Science* 15(2): 149–86. https://doi.org/10.17791/jcs.2014.15.2.149

Gamez, D. (2014b). Conscious Sensation, Conscious Perception and Sensorimotor Theories of Consciousness. In *Contemporary Sensorimotor Theory*, edited by J. M. Bishop and A. O. Martin. Switzerland: Springer International Publishing, pp. 159–74. https://doi.org/10.1007/978-3-319-05107-9_11

Gamez, D. (2014c). The measurement of consciousness: a framework for the scientific study of consciousness. *Frontiers in Psychology* 5: 714. https://doi.org/10.3389/fpsyg.2014.00714

Gamez, D. (2016). Are Information or Data Patterns Correlated with Consciousness? *Topoi* 35(1): 225–39. https://doi.org/10.1007/s11245-014-9246-7

Gamez, D. and Aleksander, I. (2011). Accuracy and performance of the state-based Φ and liveliness measures of information integration. *Consciousness and Cognition* 20(4): 1403–24. https://doi.org/10.1016/j.concog.2011.05.016

Gamez, D., Fountas, Z. and Fidjeland, A. K. (2013). A Neurally-controlled Computer Game Avatar with Human-like Behaviour. *IEEE Transactions on Computational Intelligence and AI In Games* 5(1): 1–14. https://doi.org/10.1109/tciaig.2012.2228483

Gescheider, G. A. (1997). *Psychophysics: The Fundamentals*. Mahwah, N. J.; London: Lawrence Erlbaum Associates.

Godwin, D., Barry, R. L. and Marois, R. (2015). Breakdown of the brain's functional network modularity with awareness. *Proceedings of the National Academy of Sciences of the United States of America*. https://doi.org/10.1073/pnas.1414466112

Granger, C. (1969). Investigating causal relations by econometric models and cross-spectral methods. *Econometrica* 37: 424–38. https://doi.org/10.2307/1912791

Gravato Marques, H. and Holland, O. (2009). Architectures for Functional Imagination. *Neurocomputing* 72(4–6): 743–59. https://doi.org/10.1016/j.neucom.2008.06.016

Grier, D. A. F. (2005). *When Computers were Human*. Princeton, N. J.; Oxford: Princeton University Press.

Hameroff, S. and Penrose, R. (1996). Orchestrated Reduction of Quantum Coherence in Brain Microtubules: A Model for Consciousness?. *Mathematics and Computers in Simulation* 40: 453–80. https://doi.org/10.1016/0378-4754(96)80476-9

Harary, K. and Weintraub, P. (1989). *Have an Out-of-body Experience in 30 days*. Wellingborough: Aquarian, 1990.

Harnad, S. (1994). Levels of Functional Equivalence in Reverse Bioengineering: The Darwinian Turing Test for Artificial Life. *Artificial Life* 1(3): 293–301. https://doi.org/10.1162/artl.1994.1.293

Hayes, J. E., Wallace, M. R., Knopik, V. S., Herbstman, D. M., Bartoshuk, L. M. and Duffy, V. B. (2011). Allelic variation in TAS2R bitter receptor genes associates with variation in sensations from and ingestive behaviors toward common bitter beverages in adults. *Chemical Senses* 36(3): 311–19. https://doi.org/10.1093/chemse/bjq132

Hernández-Orallo, J. and Dowe, D. L. (2010). Measuring universal intelligence: Towards an anytime intelligence test. *Artificial Intelligence* 174: 1508–39. https://doi.org/10.1016/j.artint.2010.09.006

Hibbard, B. (2011). Measuring agent intelligence via hierarchies of environments. *Proceedings of AGI 2011*, edited by J. Schmidhuber, K. R. Thórisson and M. Looks, Springer-Verlag, pp. 303–08. https://doi.org/10.1007/978-3-642-22887-2_34

Holland, O., (ed.) (2003). *Machine Consciousness*. Exeter: Imprint Academic.

Horsman, C., Stepney, S., Wagner, R. C. and Kendon, V. (2014). When does a physical system compute? *Proceedings of the Royal Society A* 470: 20140182. https://doi.org/10.1098/rspa.2014.0182

Hume, D. (1993). *An Enquiry Concerning Human Understanding*. Indianapolis: Hackett Publishing Company.

Humphreys, P. (2004). *Extending Ourselves: Computational Science, Empiricism, and Scientific Method*. Oxford: Oxford University Press.

Hurlburt, R. T. and Akhter, S. A. (2006). The Descriptive Experience Sampling method. *Phenomenology and the Cognitive Sciences* 5: 271–301. https://doi.org/10.1007/s11097-006-9024-0

Husserl, E. (1960). *Cartesian Meditations: An Introduction to Phenomenology*. Translated by D. Cairns. The Hague: Martinus Nijhoff.

Husserl, E. (1964). *The Phenomenology of Internal Time-Consciousness*. Translated by J. S. Churchill. The Hague: Martinus Nijhoff.

Huxley, A. (1965). *The Doors of Perception; Heaven and Hell*. Harmondsworth: Penguin.

Indiveri, G., Linares-Barranco, B., Hamilton, T. J., van Schaik, A., Etienne-Cummings, R., Delbruck, T., Liu, S. C., Dudek, P., Hafliger, P., Renaud, S., Schemmel, J., Cauwenberghs, G., Arthur, J., Hynna, K., Folowosele, F., Saighi, S., Serrano-Gotarredona, T., Wijekoon, J., Wang, Y. and Boahen, K. (2011). Neuromorphic silicon neuron circuits. *Frontiers in Neuroscience* 5: 73. https://doi.org/10.3389/fnins.2011.00073

Izhikevich, E. M. and Edelman, G. M. (2008). Large-scale model of mammalian thalamocortical systems. *Proceedings of the National Academy of Sciences of the United States of America* 105(9): 3593–98. https://doi.org/10.1073/pnas.0712231105

Jackendoff, R. (1987). *Consciousness and the Computational Mind.* Cambridge, Mass.; London: MIT Press in association with the British Psychological Society.

James, W. (1985). *The Varieties of Religious Experience.* Harmondsworth: Penguin.

Joy, B. (2000). Why the future doesn't need us. *Wired* 8.04.

Kaczynski, T. J. (1996). *The Unabomber Manifesto: Industrial Society and its Future.* Berkeley, California: Jolly Roger.

Kant, I. (1996). *Critique of Pure Reason.* Translated by W. S. Pluhar. Indianapolis: Hackett Publishing Company.

Kentridge, R. W. (1994). Symbols, Neurons, Soap-Bubbles and the Computation Underlying Cognition. *Minds and Machines* 4(4): 439–49. https://doi.org/10.1007/BF00974169

Kim, J. (1998). *Mind in a Physical World: An Essay on the Mind-body Problem and Mental Causation.* Cambridge, Mass.: MIT Press.

Knott, G., Marchman, H., Wall, D. and Lich, B. (2008). Serial section scanning electron microscopy of adult brain tissue using focused ion beam milling. *Journal of Neuroscience* 28(12): 2959–64. https://doi.org/10.1523/JNEUROSCI.3189-07.2008

Koch, C. (2004). *The Quest for Consciousness: A Neurobiological Approach.* Englewood: Roberts & Company.

Koch, C., Massimini, M., Boly, M. and Tononi, G. (2016). Neural correlates of consciousness: progress and problems. *Nature Reviews Neuroscience* 17(5): 307–21. https://doi.org/10.1038/nrn.2016.22

Kotze, H. F. and Moller, A. T. (1990). Effect of auditory subliminal stimulation on GSR. *Psychology Reports* 67(3 Pt 1): 931–34.

Kouider, S. and Dehaene, S. (2007). Levels of processing during non-conscious perception: a critical review of visual masking. *Philosophical Transactions of the Royal Society B: Biological Sciences* 362(1481): 857–75. https://doi.org/10.1098/rstb.2007.2093

Kouider, S., Stahlhut, C., Gelskov, S. V., Barbosa, L. S., Dutat, M., de Gardelle, V., Christophe, A., Dehaene, S. and Dehaene-Lambertz, G. (2013). A neural marker of perceptual consciousness in infants. *Science* 340(6130): 376–80. https://doi.org/10.1126/science.1232509

Kuhn, T. S. (1962). *The Structure of Scientific Revolutions.* Chicago, London: University of Chicago Press.

Lamme, V. A. (2006). Towards a true neural stance on consciousness. *Trends in Cognitive Sciences* 10(11): 494–501. https://doi.org/10.1016/j.tics.2006.09.001

Lamme, V. A. F. (2010). How neuroscience will change our view on consciousness. *Cognitive Neuroscience* 1(3): 204–40. https://doi.org/10.1080/17588921003731586

Latour, B. (1987). *Science in Action: How to Follow Scientists and Engineers through Society*. Milton Keynes: Open University Press.

Laureys, S., Antoine, S., Boly, M., Elincx, S., Faymonville, M. E., Berre, J., Sadzot, B., Ferring, M., De Tiege, X., van Bogaert, P., Hansen, I., Damas, P., Mavroudakis, N., Lambermont, B., Del Fiore, G., Aerts, J., Degueldre, C., Phillips, C., Franck, G., Vincent, J. L., Lamy, M., Luxen, A., Moonen, G., Goldman, S. and Maquet, P. (2002). Brain function in the vegetative state. *Acta Neurologica Belgica* 102(4): 177–85.

Lee, U., Mashour, G. A., Kim, S., Noh, G. J. and Choi, B. M. (2009). Propofol induction reduces the capacity for neural information integration: implications for the mechanism of consciousness and general anesthesia. *Consciousness and Cognition* 18(1): 56–64. https://doi.org/10.1016/j.concog.2008.10.005

Legg, S. and Hutter, M. (2007). Universal intelligence: A definition of machine intelligence. *Minds and Machines* 17: 391–444. https://doi.org/10.1007/s11023-007-9079-x

Legon, W., Sato, T. F., Opitz, A., Mueller, J., Barbour, A., Williams, A. and Tyler, W. J. (2014). Transcranial focused ultrasound modulates the activity of primary somatosensory cortex in humans. *Nature Neuroscience* 17(2): 322–29. https://doi.org/10.1038/nn.3620

Lehar, S. (2003). *The World in Your Head*. Mahwah, New Jersey: Lawrence Erlbaum Associates.

Lewis, D. (1973). Causation. *Journal of Philosophy* 70(17): 556–67.

Libet, B. (1985). Unconscious cerebral initiative and the role of conscious will in voluntary action. *Behavioral and Brain Sciences* 6(8–9): 47–57. https://doi.org/10.1017/S0140525X00044903

Locke, J. (1997). *An Essay Concerning Human Understanding*. London: Penguin Books.

Lucretius Carus, T. (2007). *The Nature of Things*. Translated by A. E. Stallings. London: Penguin.

Lutz, A., Lachaux, J. P., Martinerie, J. and Varela, F. J. (2002). Guiding the study of brain dynamics by using first-person data: synchrony patterns correlate with ongoing conscious states during a simple visual task. *Proceedings of the National Academy of Sciences of the United States of America* 99(3): 1586–91. https://doi.org/10.1073/pnas.032658199

Mackie, J., L. (1993). Causes and Conditions. In *Causation*, edited by E. Sosa and M. Tooley. Oxford: Oxford University Press, pp. 33–55.

Mainland, J. D., Keller, A., Li, Y. R., Zhou, T., Trimmer, C., Snyder, L. L., Moberly, A. H., Adipietro, K. A., Liu, W. L., Zhuang, H., Zhan, S., Lee, S. S., Lin, A. and Matsunami, H. (2014). The missense of smell: functional variability in the human odorant receptor repertoire. *Nature Neuroscience* 17(1): 114–20. https://doi.org/10.1038/nn.3598

Markram, H. (2006). The Blue Brain project. *Nature Reviews Neuroscience* 7(2): 153–60. https://doi.org/10.1038/nrn1848

Marx, V. (2014). Neurobiology: rethinking the electrode. *Nature Methods* 11: 1099–103. https://doi.org/10.1038/nmeth.3149

Massimini, M., Boly, M., Casali, A., Rosanova, M. and Tononi, G. (2009). A perturbational approach for evaluating the brain's capacity for consciousness. *Progress in Brain Research* 177: 201–14. https://doi.org/10.1016/S0079-6123(09)17714-2

Mayerich, D., Abbott, L. and McCormick, B. (2008). Knife-edge scanning microscopy for imaging and reconstruction of three-dimensional anatomical structures of the mouse brain. *Journal of Microscopy* 231(Pt 1): 134–43. https://doi.org/10.1111/j.1365-2818.2008.02024.x

McFadden, J. (2002). Synchronous Firing and its Influence on the Brain's Electromagnetic Field. *Journal of Consciousness Studies* 9(4): 23–50.

McGinn, C. (1989). Can We Solve the Mind-Body Problem? *Mind* 98(391): 349–66.

Melzack, R. (1990). Phantom limbs and the concept of a neuromatrix. *Trends in Neurosciences* 13(3): 88–92.

Melzack, R. (1992). Phantom limbs. *Scientific American* 266(4): 120–26.

Merikle, P. M. and Daneman, M. (1996). Memory for unconsciously perceived events: evidence from anesthetized patients. *Consciousness and Cognition* 5(4): 525–41. https://doi.org/10.1006/ccog.1996.0031

Metzinger, T. (2000). Consciousness Research at the End of the Twentieth Century. In *Neural Correlates of Consciousness: Empirical and Conceptual Questions*, edited by T. Metzinger. Cambridge, Mass.: MIT Press, pp. 1–12.

Metzinger, T. (2003). *Being No One: The Self-model Theory of Subjectivity*. Cambridge, Mass.: MIT Press.

Mnih, V., Kavukcuoglu, K., Silver, D., Rusu, A. A., Veness, J., Bellemare, M. G., Graves, A., Riedmiller, M., Fidjeland, A. K., Ostrovski, G., Petersen, S., Beattie, C., Sadik, A., Antonoglou, I., King, H., Kumaran, D., Wierstra, D., Legg, S. and Hassabis, D. (2015). Human-level control through deep reinforcement learning. *Nature* 518(7540): 529–33. https://doi.org/10.1038/nature14236

Monti, M. M., Vanhaudenhuyse, A., Coleman, M. R., Boly, M., Pickard, J. D., Tshibanda, L., Owen, A. M. and Laureys, S. (2010). Willful modulation of brain activity in disorders of consciousness. *New England Journal of Medicine* 362(7): 579–89. https://doi.org/10.1056/NEJMoa0905370

Moor, J. H. (1988). Testing Robots for Qualia. In *Perspectives on Mind*, edited by H. R. Otto and J. A. Tuedio. Dordrecht/Boston/Lancaster/Tokyo: D. Reidel Publishing Company.

Moravec, H. (1988). *Mind Children: The Future of Robot and Human Intelligence.* Cambridge, Massachusetts; London: Harvard University Press.

Nagel, S. K., Carl, C., Kringe, T., Martin, R. and Konig, P. (2005). Beyond sensory substitution—learning the sixth sense. *Journal of Neural Engineering* 2(4): R13–26. https://doi.org/10.1088/1741-2560/2/4/R02

Nagel, T. (1974). What Is It Like to Be a Bat? *The Philosophical Review* 83: 435–56.

Nikolenko, V., Poskanzer, K. E. and Yuste, R. (2007). Two-photon photostimulation and imaging of neural circuits. *Nature Methods* 4(11): 943–50. https://doi.org/10.1038/nmeth1105

Nishimoto, S., Vu, A. T., Naselaris, T., Benjamini, Y., Yu, B. and Gallant, J. L. (2011). Reconstructing visual experiences from brain activity evoked by natural movies. *Current Biology* 21(19): 1641–46. https://doi.org/10.1016/j.cub.2011.08.031

Noë, A. (2004). *Action in Perception.* Cambridge, Massachusetts; London: MIT.

O'Regan, J. K. (1992). Solving the "real" mysteries of visual perception: the world as an outside memory. *Canadian Journal of Psychology* 46(3): 461–88.

O'Regan, J. K. and Noë, A. (2001). A sensorimotor account of vision and visual consciousness. *Behavioral and Brain Sciences* 24(5): 939–73. https://doi.org/10.1017/S0140525X01000115

Öhman, A. and Soares, J. J. F. (1994). "Unconscious Anxiety": Phobic Responses to Masked Stimuli. *Journal of Abnormal Psychology* 103(2): 231–40.

Oizumi, M., Albantakis, L. and Tononi, G. (2014). From the phenomenology to the mechanisms of consciousness: Integrated Information Theory 3.0. *PLoS Computational Biology* 10(5): e1003588. https://doi.org/10.1371/journal.pcbi.1003588

Ophiel (1970). *The Art and Practice of Astral Projection (Ninth Edition).* St. Paul, Minnesota: Peach Publishing Co.

Owen, A. M., Coleman, M. R., Boly, M., Davis, M. H., Laureys, S. and Pickard, J. D. (2006). Detecting awareness in the vegetative state. *Science* 313(5792): 1402. https://doi.org/10.1126/science.1130197

Palmer, S. E. (1999). Color, consciousness, and the isomorphism constraint. *Behavioral and Brain Sciences* 22(6): 923–43.

Parker, E. S., Cahill, L. and McGaugh, J. L. (2006). A case of unusual autobiographical remembering. *Neurocase* 12(1): 35–49. https://doi.org/10.1080/13554790500473680

Pedhazur, E. J. and Schmelkin, L. P. (1991). *Measurement, design, and analysis: an integrated approach.* Hillsdale, N. J.: Lawrence Erlbaum Associates.

Persaud, N., McLeod, P. and Cowey, A. (2007). Post-decision wagering objectively measures awareness. *Nature Neuroscience* 10(2): 257–61. https://doi.org/10.1038/nn1840

Petitmengin, C. (2006). Describing one's subjective experience in the second person: An interview method for the science of consciousness. *Phenomenology and the Cognitive Sciences* 5: 229–69. https://doi.org/10.1007/s11097-006-9022-2

Piccinini, G. (2007). Computing Mechanisms. *Philosophy of Science* 74: 501–26. https://doi.org/10.1086/522851

Piccinini, G. (2012) Computation in Physical Systems. *The Stanford Encyclopedia of Philosophy (Fall 2012 Edition)*, edited by. E. N. Zalta. http://plato.stanford.edu/archives/fall2012/entries/computation-physicalsystems/

Pitts, M. A., Metzler, S. and Hillyard, S. A. (2014). Isolating neural correlates of conscious perception from neural correlates of reporting one's perception. *Frontiers in Psychology* 5: 1078. https://doi.org/10.3389/fpsyg.2014.01078

Pockett, S. (2000). *The Nature of Consciousness: A Hypothesis*. San Jose, California: Writers Club Press.

Popper, K. R. (2002). *The Logic of Scientific Discovery*. London: Routledge.

Prinz, J. J. (2003). Level-headed Mysterianism and artificial experience. In *Machine Consciousness*, edited by O. Holland. Exeter: Imprint Academic.

Putnam, H. (1975). *Mind, Language and Reality*. Cambridge: Cambridge University Press.

Putnam, H. (1988). *Representation and Reality*. Cambridge, Massachusetts; London: MIT Press.

Ramirez, S., Liu, X., Lin, P. A., Suh, J., Pignatelli, M., Redondo, R. L., Ryan, T. J. and Tonegawa, S. (2013). Creating a false memory in the hippocampus. *Science* 341(6144): 387–91. https://doi.org/10.1126/science.1239073

Rast, A., Galluppi, F., Davies, S., Plana, L., Patterson, C., Sharp, T., Lester, D. and Furber, S. (2011). Concurrent heterogeneous neural model simulation on real-time neuromimetic hardware. *Neural Networks* 24(9): 961–78. https://doi.org/10.1016/j.neunet.2011.06.014

Rees, G., Kreiman, G. and Koch, C. (2002). Neural correlates of consciousness in humans. *Nature Reviews Neuroscience* 3(4): 261–70. https://doi.org/10.1038/nrn783

Reggia, J. A. (2013). The rise of machine consciousness: studying consciousness with computational models. *Neural Networks* 44: 112–31. https://doi.org/10.1016/j.neunet.2013.03.011

Rensink, R. A., O'Regan, J. K. and Clark, J. J. (1997). To See or Not to See: The Need for Attention to Perceive Changes in Scenes. *Psychological Science* 8(5): 368–73. https://doi.org/10.1111/j.1467-9280.1997.tb00427.x

Rorty, R. (1980). *Philosophy and the Mirror of Nature*. Oxford: Basil Blackwell.

Russell, B. (1927). *An Outline of Philosophy*. London: G. Allen & Unwin.

Sandberg, K., Timmermans, B., Overgaard, M. and Cleeremans, A. (2010). Measuring consciousness: is one measure better than the other? *Consciousness and Cognition* 19(4): 1069–78. https://doi.org/10.1016/j.concog.2009.12.013

Schmidt, M. and Lipson, H. (2009). Distilling free-form natural laws from experimental data. *Science* 324(5923): 81–85. https://doi.org/10.1126/science.1165893

Schreiber, T. (2000). Measuring information transfer. *Physical Review Letters* 85(2): 461–64. https://doi.org/10.1103/PhysRevLett.85.461

Schule, M. (2014). Natural Computation: the Cellular Automata Case. *Proceedings of 7th AISB Symposium on Computing and Philosophy*, edited by J. Preston and Y. Erden, London, AISB.

Schutz, K., Schendzielarz, I., Zwitserlood, P. and Vorberg, D. (2007). Nice wor_ if you can get the wor_: subliminal semantic and form priming in fragment completion. *Consciousness and Cognition* 16(2): 520–32. https://doi.org/10.1016/j.concog.2006.09.001

Searle, J. R. (1980). Minds, Brains, and Programs. *Behavioral and Brain Sciences* 3(3): 417–57. https://doi.org/10.1017/S0140525X00005756

Seo, D., Carmena, J. M., Rabaey, J. M., Alon, E. and Maharbiz, M. M. (2013). Neural Dust: An Ultrasonic, Low Power Solution for Chronic Brain-Machine Interfaces. *arXiv* 1307.2196.

Seth, A. K. (2009). The Strength of Weak Artificial Consciousness. *International Journal of Machine Consciousness* 1(1): 71–82. https://doi.org/10.1142/S1793843009000086

Seth, A. K., Dienes, Z., Cleeremans, A., Overgaard, M. and Pessoa, L. (2008). Measuring consciousness: relating behavioural and neurophysiological approaches. *Trends in Cognitive Sciences* 12(8): 314–21. https://doi.org/10.1016/j.tics.2008.04.008

Seth, A. K., Izhikevich, E., Reeke, G. N. and Edelman, G. M. (2006). Theories and measures of consciousness: an extended framework. *Proceedings of the National Academy of Sciences of the United States of America* 103(28): 10799–804. https://doi.org/10.1073/pnas.0604347103

Shagrir, O. (2005). The Rise and Fall of Computational Functionalism. In *Hilary Putnam*, edited by Y. Ben-Menahem. Cambridge: Cambridge University Press, pp. 220–50.

Shanahan, M. (2008). A spiking neuron model of cortical broadcast and competition. *Consciousness and Cognition* 17(1): 288–303. https://doi.org/10.1016/j.concog.2006.12.005

Shanahan, M. (2010). *Embodiment and the Inner Life: Cognition and Consciousness in the Space of Possible Minds*. Oxford: Oxford University Press.

Shanahan, M. and Wildie, M. (2012). Knotty-centrality: finding the connective core of a complex network. *PLoS One* 7(5): e36579. https://doi.org/10.1371/journal.pone.0036579

Shannon, C. E. (1948). A Mathematical Theory of Communication. *The Bell System Technical Journal* 27: 379–423, 623–56.

Simons, D. J. and Chabris, C. F. (1999). Gorillas in our midst: sustained inattentional blindness for dynamic events. *Perception* 28(9): 1059–74. https://doi.org/10.1068/p281059

Simons, D. J. and Rensink, R. A. (2005). Change blindness: past, present, and future. *Trends in Cognitive Sciences* 9(1): 16–20. https://doi.org/10.1016/j.tics.2004.11.006

Sloman, A. (2006). Why Asimov's Three Laws of Robotics Are Unethical. http://www.cs.bham.ac.uk/research/projects/cogaff/misc/asimov-three-laws.html

Smart, J. J. C. (1959). Sensations and Brain Processes. *The Philosophical Review* 68(2): 141–56.

Sparkes, A., Aubrey, W., Byrne, E., Clare, A., Khan, M. N., Liakata, M., Markham, M., Rowland, J., Soldatova, L. N., Whelan, K. E., Young, M. and King, R. D. (2010). Towards Robot Scientists for autonomous scientific discovery. *Automated Experimentation* 2: 1. https://doi.org/10.1186/1759-4499-2-1

Stevens, S. S. (1968). Measurement, statistics, and the schemapiric view. Like the faces of Janus, science looks two ways—toward schematics and empirics. *Science* 161(3844): 849–56.

Strawson, G. (2015). The Consciousness Myth. *The Times Literary Supplement* 5839: 14–15.

Taylor, C. C. W. (1999). *The Atomists: Leucippus and Democritus: Fragments: A Text and Translation with a Commentary.* Toronto: University of Toronto Press.

Teasdale, G. and Jennett, B. (1974). Assessment of coma and impaired consciousness. A practical scale. *Lancet* 2(7872): 81–84.

Tononi, G. (2004). An information integration theory of consciousness. *BMC Neuroscience* 5: 42. https://doi.org/10.1186/1471-2202-5-42

Tononi, G. (2008). Consciousness as Integrated Information: A Provisional Manifesto. *Biological Bulletin* 215(3): 216–42. https://doi.org/10.2307/25470707

Tononi, G. (2012). *Phi: a voyage from the brain to the soul.* New York: Pantheon.

Tononi, G. and Koch, C. (2008). The neural correlates of consciousness: an update. *Annals of the New York Academy of Sciences* 1124: 239–61. https://doi.org/10.1196/annals.1440.004

Tononi, G. and Sporns, O. (2003). Measuring information integration. *BMC Neuroscience* 4: 31. https://doi.org/10.1186/1471-2202-4-31

van der Hoort, B., Guterstam, A. and Ehrsson, H. H. (2011). Being Barbie: the size of one's own body determines the perceived size of the world. *PLoS One* 6(5): e20195. https://doi.org/10.1371/journal.pone.0020195

Van Heuveln, B., Dietrich, E. and Oshima, M. (1998). Let's dance! The equivocation in Chalmers' dancing qualia argument. *Minds and Machines* 8: 237–49. https://doi.org/10.1023/A:1008273402702

van Rijn, C. M., Krijnen, H., Menting-Hermeling, S. and Coenen, A. M. (2011). Decapitation in rats: latency to unconsciousness and the 'wave of death'. *PLoS One* 6(1): e16514. https://doi.org/10.1371/journal.pone.0016514

VanRullen, R. and Koch, C. (2003). Is perception discrete or continuous? *Trends in Cognitive Sciences* 7(5): 207–13. https://doi.org/10.1016/S1364-6613(03)00095-0

Vigen, T. (2016). Spurious Correlations. http://tylervigen.com/spurious-correlations

Warwick, K., Xydas, D., Nasuto, S. J., Becerra, V. M., Hammond, M. W., Downes, J. H., Marshall, S. and Whalley, B. J. (2010). Controlling a Mobile Robot with a Biological Brain. *Defence Science Journal* 60(1): 5–14.

Wegner, D. M. (2002). *The Illusion of Conscious Will*. Cambridge, Mass.; London: MIT Press.

Weiskrantz, L. (1986). *Blindsight: A Case Study and Implications*. Oxford: Clarendon.

Wheeler, J. A. (1990). Information, Physics, Quantum: The Search for Links. In *Complexity, Entropy, and the Physics of Information*, edited by W. Zurek. Redwood City: Addison-Wesley.

Wilkes, K. V. (1988a). *Real People: Personal Identity without Thought Experiments*. Oxford: Clarendon.

Wilkes, K. V. (1988b). ___, yìshì, duh, um, and consciousness. In *Consciousness in Contemporary Science*, edited by A. J. Marcel and E. Bisiach. Oxford: Clarendon Press, pp. 16–41.

Wilkins, L. K., Girard, T. A. and Cheyne, J. A. (2011). Ketamine as a primary predictor of out-of-body experiences associated with multiple substance use. *Consciousness and Cognition* 20: 943–50. https://doi.org/10.1016/j.concog.2011.01.005

Wilson, D. L. (1993). Quantum theory and consciousness. *Behavioral and Brain Sciences* 16(3): 615–16. https://doi.org/10.1017/S0140525X00031952

Wilson, D. L. (1999). Mind-Brain Interaction and the Violation of Physical Laws. *Journal of Consciousness Studies* 6(8–9): 185–200.

Wittgenstein, L. (1969). *On Certainty*. Translated by D. Paul and G. E. M. Anscombe. Oxford: Blackwell.

Wolfram, S. (2002). *A New Kind of Science*. Champaign, Ill.: Wolfram Media; London: Turnaround.

Yang, B., Treweek, J. B., Kulkarni, R. P., Deverman, B. E., Chen, C. K., Lubeck, E., Shah, S., Cai, L. and Gradinaru, V. (2014). Single-cell phenotyping within transparent intact tissue through whole-body clearing. *Cell* 158(4): 945–58. https://doi.org/10.1016/j.cell.2014.07.017

Yin, M., Borton, D. A., Aceros, J., Patterson, W. R. and Nurmikko, A. V. (2013). A 100-channel hermetically sealed implantable device for chronic wireless neurosensing applications. *IEEE Transactions on Biomedical Circuits and Systems* 7(2): 115–28. https://doi.org/10.1109/TBCAS.2013.2255874

Zeki, S. and Bartels, A. (1999). Towards a theory of visual consciousness. *Consciousness and Cognition* 8(2): 225–59. https://doi.org/10.1006/ccog.1999.0390

Zingg, B., Hintiryan, H., Gou, L., Song, M. Y., Bay, M., Bienkowski, M. S., Foster, N. N., Yamashita, S., Bowman, I., Toga, A. W. and Dong, H. W. (2014). Neural networks of the mouse neocortex. *Cell* 156(5): 1096–111. https://doi.org/10.1016/j.cell.2014.02.023

Zuse, K. (1970). *Calculating Space*. Cambridge, Mass.: Massachusetts Institute of Technology, Project MAC.

Zylberberg, A., Fernandez Slezak, D., Roelfsema, P. R., Dehaene, S. and Sigman, M. (2010). The brain's router: a cortical network model of serial processing in the primate brain. *PLoS Computational Biology* 6(4): e1000765. https://doi.org/10.1371/journal.pcbi.1000765

Index

access consciousness. *See* phenomenal and access consciousness
anaesthetics 173
animal consciousness 117, 121, 130, 173
artificial consciousness. *See* machine consciousness
artificial intelligence 83, 123, 142
assumptions for the measurement of consciousness 3, 48, 49, 52, 53, 55, 58, 59, 60, 151
atomism 15, 18, 31

bat consciousness 117, 118, 122
behaviour associated with consciousness. *See* conscious behaviour
brain damage 126, 140, 173
brain in a vat 55
brain measurement technology 152
brain modification technology 130
brute regularity **41**, 81, 169
bubble of experience **12**, 126, 155
bubble of perception 11

calculator 103, 108
Cartesian theatre 165
causation 56
CC set **54**, 74, 80, 86, 89, 90, 108, 110, 116, 129, 130, 136, 137, 141
c-description **65**, 115, 119, 129, 132, 137, 139
colour inversion 50
coma patients 117, 126
computational discovery of theories of consciousness 83
computation c-theory **106**
computer 93, 103, 104, 109
concept of consciousness **26**, 167, 176

conscious behaviour 44, 170
consciousness, definition of. *See* definition of consciousness
conservative deduction **121**, 131
constraints on theories of consciousness 73
contents of consciousness 27, 126
context 71. *See also* physical context
correlates of consciousness 73, 74, 79. *See also* CC set
c-report **45**, 48, 49, 56, 59, 73, 86, 101, 118, 141, 175, 176
c-theory **79**, 82, 113, 115, 116, 119, 129, 137, 139, 153
custom interface 95, 99

data 95
deduction **118**. *See also* conservative deduction, liberal deduction
definition of consciousness 26
description of consciousness. *See* c-description
description of physical world. *See* p-description
dreams 11, 60
drug-induced experiences 11, 126
dualism 151

e-causation **57**, **58**, 73, 86, 99
effective connectivity 60, 87, 153, 168
enhancement of consciousness 126, 129
epiphenomenalism 151
ethical issues 119, 142, 144, 146
existential threat to humanity 142
experiments on consciousness 48, 49, 61, 63, 74, 76, 88, 98, 100, 111, 113, 115, 119, 150, 170
eye movements 61

first-person report 43, 50, 176. *See also* c-report
functional connectivity 48, 55, 87, 168, 174
functionalism. *See* computation c-theory

general-purpose computer 104
global workspace theory 87

hallucination 13
hard problem of consciousness 36, 38, 41, 149

imagination 34, 36, 38, 52, 81, 82, 118, 154
implantation of chip in brain 77, 101, 131, 140, 179
implementation of computation 110
information **93**, **94**
information c-theory **97**
information entropy 96
information integration theory of consciousness 65, 87, 98
information processing 93, **109**, **110**
interface **94**, 97, 109
intrinsic property 73, 86, 100, 178
inverted qualia. *See* colour inversion
invisible physical world 13, 16, **22**, 26, 37, 69

language. *See* natural language
level of consciousness 29, 126, 167
liberal deduction **121**, 132
limitations of this interpretation of consciousness 54, 64, 78, 117, 119, 140, 152, 153

machine consciousness 72, 117, 121, 123, **135**
material **85**, **86**, 89, 101, 121, 140
MC1 machine consciousness 136, 142, 144
MC2 machine consciousness 136
MC3 machine consciousness 136
MC4 machine consciousness **136**, 139, 140, 144, 146

measurement of consciousness 61. *See also* c-report
measurement of physical world **69**, 152
memory 62, 127
modification of consciousness **125**, 129
mystical experience 127, 132

naive realism 10, 11, 26
natural experiment **76**, 101, 131
natural language 62, 65, 71, 80, 151
nc-report 46
neural correlates of consciousness 72, 87. *See also* correlates of consciousness
neuromorphic chips 138, 141
neuron, definition of 71

offline bubble of experience 13, 20, 29, 35
online bubble of experience 12, 20, 29, 35
out-of-body experience 126

panpsychism 50, 111, 151
p-description **72**, 115, 119, 129, 137, 139, 152
perception 11, 18, 87, 127, 165, 172, 192
phenomenal and access consciousness 48, 171
phenomenology 33, 64, 136, 155
philosophical problems with consciousness 3, 151
philosophy of science 181
physical context **121**, 131
physical c-theory **85**, 98, 100, 113, 151
physicalism 31, 33, 59, 151, 154
physical world 81, 85
platinum standard system **47**, **53**, 76, 77, 116, 117, 140, 141, 154, 173
prediction 80, **113**, 115, 116
primary qualities 17, 20, 22

qualia 9

reduction of consciousness to
 physical world. *See* physicalism
robots 71, 117, 142, 143

saccade. *See* eye movements
scientific experiments.
 See experiments on consciousness
secondary qualities 17, 20, 22
sensorimotor theory 20, 166
simulation of brain 83, 141, 153, 195
singularity 143
solipsism 47
space 127, 166
special-purpose computer 103
specious present 176
structural connectivity 168

subjectivity of computation 108
subjectivity of information 95, 98
supervenience 52, 55, 172

testable prediction. *See* prediction
theory of consciousness. *See* c-theory
time 61, 127, 166
Turing machine 104

unnatural experiment **76**
uploading consciousness 141

virtual reality 114

wakefulness 167

zombies 47, 145

www.ingramcontent.com/pod-product-compliance
Lightning Source LLC
Chambersburg PA
CBHW060603230426
43670CB00011B/1942